Seamless Assessment
in Science

Seamless
Assessment
in Science

A Guide for Elementary and Middle School Teachers

Sandra K. Abell and Mark J. Volkmann

Science Education Center

University of Missouri–Columbia

With contributions from colleagues in
Columbia Public Schools and at the
University of Missouri–Columbia

Heinemann, *Portsmouth, NH*

NSTA, *Arlington, VA*

Heinemann

361 Hanover Street
Portsmouth, NH 03801–3912
www.heinemann.com

National Science Teachers Association
1840 Wilson Boulevard
Arlington, VA 22201-3000, USA
www.nsta.org

Offices and agents throughout the world

Library of Congress Cataloging-in-Publication Data
Abell, Sandra K.
    Seamless assessment in science : a guide for elementary and middle school teachers /
Sandra K. Abell and Mark J. Volkmann.
        p. cm.
    Includes bibliographical references and appendix.
    ISBN 0-325-00769-1 (alk. paper)
    1. Science—Study and teaching (Elementary)—United States—Evaluation.
2. Science—Study and teaching (Secondary)—United States—Evaluation.   3. Science—
Ability testing—United States.   I. Volkmann, Mark J.   II. Title.

LB1585.3.A24 2006
372.3'504—dc22   2005031620

Editor: Robin Najar
Production: Patricia Adams
Typesetter: SPI Publisher Services
Cover design: Jenny Jensen Greenleaf
Manufacturing: Louise Richardson

Printed in the United States of America on acid-free paper
10   09   RRD   4   5

To our son, Luke. Teachers in your house at all hours. Endless pizza dinners. Parents so busy writing they forget the school pickup. You have been patient with us and kind to our visitors. We love you very much.

Contents

Foreword

Teachers of science face the incredibly complex task of combining what we know about how students learn, science content that in itself is complicated and often difficult to understand, and scientific inquiry. Then, as if this were not enough, they have to assess the results of their teaching—student understanding. Reasons for avoiding science in the elementary school are not hard to come by. It would be nice if elementary—and for that matter, middle school—teachers had some helpful assistance through this educational maze. Well, they do. This little book, *Seamless Assessment in Science*, provides practical suggestions that will help all elementary and middle school teachers, both preservice and inservice. Let me say more about some of the book's features, especially the 5E instructional model.

In the late 1980s my colleagues and I at Biological Sciences Curriculum Study (BSCS) received funding to develop a new program for elementary school science. As we began working, we confronted a fundamental problem— how could we incorporate research on learning in a way that was helpful for teachers? In particular, we had to balance what researchers described as a constructivist model of learning and the classroom constraints that elementary and middle school teachers face. Further, we had to create a program that teachers could comprehend and apply in order to achieve positive results in student learning. In short, the model had to be understandable, manageable, and practical. We adapted a model originally developed by Bob Karplus and colleagues and used it in the new BSCS program. We realized the model worked, so we also used it in BSCS programs for middle and high school. That said, I could not have imagined the widespread use of the BSCS 5E model two decades later. The model has been used in state frameworks, other science programs, and now in this guidebook for elementary and middle school teachers.

At the time of our original work, I maintained that the 5E instructional model was a necessary but not sufficient condition to bring about conceptual change within the structure of school science programs. The model provided adequate time and appropriate opportunities for conceptual change. But there remained the critical interaction between teacher and student. The burden of student learning is too heavy to place exclusively on an instructional model. The teacher remains the essential link between instructional materials and student learning. All of this said, Sandra Abell and Mark Volkmann have provided a wonderful extension to the 5E model. They have added an assessment component that complements each phase of the model and helps teachers identify their students' current science ideas (engage), determine how students are building understanding (explore), review students' new understanding (explain), demonstrate their ability to apply new understandings (elaborate), and determine what students have learned from lessons (evaluate). This seamless use of assessment definitely enhances the 5E model. As for the criteria stated earlier, the model is still understandable, manageable, and practical. But now it is better!

In addition to practical assessment strategies, *Seamless Assessment in Science* incorporates recent research on how students learn. The book provides an excellent bridge between research and practice. Further, it also does this for the important area of scientific inquiry, as described in the *National Science Education Standards*.

In this era of assessment, all of us have learned that assessment is more than a test. *Seamless Assessment in Science* is a sophisticated and practical resource for all elementary and middle school teachers. It is the kind of book one reads and thinks, "I wish I would have done that." Well, Sandra Abell and Mark Volkmann did, and both elementary and middle school teachers and their students will be better because they did.

Rodger W. Bybee
Executive Director
BSCS
Colorado Springs, Colorado

Preface

This book grew out of our experiences as science teachers and science teacher educators. While teaching science, we have struggled with knowing how our students are thinking and how to help them understand science better. The idea of seamless assessment took seed as we worked through these struggles alone in our classrooms. A few years ago the seed sprouted when we together recognized the potential of linking science assessment with science teaching through an instructional sequence called the 5E model. This link helped us see assessment in a new light and helped us build assessment purposefully into our instruction.

As the seed, firmly rooted in a well-accepted instructional model, started to grow, we thought about how to communicate our ideas beyond our own classrooms. In 2003, we presented our ideas about seamless assessment at the National Science Teachers Association (NSTA) national conference in Philadelphia. The participants in that session responded enthusiastically and prompted us to disseminate our ideas. Our seedling idea of seamless assessment was taking on more leaves and becoming sturdier. Thanks to encouragement from Chris Ohana, editor of NSTA's *Science and Children* journal, who attended our NSTA session, we translated our conference paper into a published article (Volkmann and Abell 2003).

Not long after the article appeared in press, Robin Najar from Heinemann contacted us about the possibility of turning our ideas into a book. We thank Robin for having the vision to grow our seedling into a tree. Her confidence in us provided the motivation that led our tree to bear fruit.

However, it is not easy to nurture a growing plant. We realized that input from classroom teachers would be essential enrichment. Thus, we found a group of local teachers who were willing to cultivate seamless assessment in their own classrooms. We collaborated with them, listened to their stories of science teaching, learning, and assessment, and integrated their stories into this book. We cannot begin to express our admiration and appreciation of these teachers. Their most important job is helping students learn, and they spend countless hours in the pursuit of excellence in science education. Yet they took time to write their stories, from which others can learn.

Our goal in writing this book is to help preservice and practicing elementary and middle school teachers of science think about science teaching, learning, and assessment as a seamless act. Chapter 1 sets the stage by outlining what counts as learning and what the standards say about assessment. In Chapter 2, we describe the 5E model of science instruction and introduce our notion of seamless assessment, which is linked to that model. We also share a variety of strategies for enacting seamless assessment. In Chapters 3, 4, and 5 we present examples of classrooms engaged in inquiry-based science instruction and seamless assessment. We have organized these chapters by science discipline. Chapter 3 focuses on life science examples, Chapter 4 on the physical sciences, and Chapter 5 on earth and space science. In each chapter, we include vignettes of science classrooms at the primary (K–3), intermediate (4–5), and middle school (6–8) levels. These vignettes, written by us and by our elementary and middle school teacher collaborators, tell stories of teaching, learning, and assessment in science classrooms. We believe that these examples, grounded in the real world of classroom teaching, will help others plan, enact, and learn from seamless assessment in science education. In the final chapter of the book, we present a summary of the seamless assessment model and what we learned by both employing it in our science teaching and using it to organize writing about our experiences. We conclude with a challenge to all teachers to incorporate inquiry-based instruction and seamless assessment into elementary and middle school science classrooms.

Our greatest hope is that readers of this volume will find the seeds of seamless assessment taking root in their own science instruction. Our ultimate goal is to grow a sustainable forest in which all students learn science.

Sandra Abell and Mark Volkmann
Columbia, Missouri

Seamless Assessment: An Introduction

1

W e teach in an era when standardized testing is hailed as the solution to many of our educational problems. Yet, at the same time, we hear teachers say that standardized tests constrain what and how they teach science, and that such tests provide little useful information about student learning. This book is about a different kind of educational assessment—the kind that takes place each day in classrooms and provides teachers with ongoing feedback about their students' science learning. We call it *seamless assessment*, because it is often indistinguishable from classroom instruction.

Seamless assessment is embedded in teaching. It is tied to instructional goals. It is a purposely planned component of the instructional sequence. It is carried out as part of science activities. It provides information to guide instruction and helps students think about their own learning. Others have written about and provided ideas for embedding assessment in science teaching (Doran et al. 2002; Hein and Price 1994; Wilson and Scalise 2003). Our approach is consistent with their ideas, but at the same time it is unique. Seamless assessment uses the 5E model (Bybee 1997; 2002), a teaching sequence for inquiry-based science instruction, to explore various purposes and strategies for classroom assessment. In the next chapter, we describe the 5E model and illustrate how it provides an organizer for seamless assessment.

Before we describe seamless assessment throughout the 5E model, we need to agree about what learning in science looks like. Once we establish that foundation, we can think about the kinds of assessment data that would provide evidence for science learning. We must also link our thinking about learning and assessment to what national standards documents have to say. In the following sections, we discuss what counts as learning and what the standards say. Then, we summarize a few key terms in the assessment field that will help communicate seamless assessment as a viable process in science teaching. Finally, we provide an overview for the rest of the book.

■ What Counts as Learning in Science

According to science assessment specialist Paul Black (2003), assessment serves three main purposes: accountability, certification, and learning. Assessment data are used to provide accountability information about teachers, schools, and districts to parents, community, and policy-making groups. Assessment is also used to certify students for further education or employment. In both of these purposes, narrow views of what counts as learning lead to assessment that is often limited to formal standardized tests. The third purpose of assessment, learning, is what concerns us in this book. The learning purpose helps shape a different kind of assessment, seamless assessment, as a tool for helping both students and teachers learn. Seamless assessment provides teachers with feedback on student learning to guide instruction; by engaging in seamless assessment tasks, learners progress.

If learning is the goal of seamless assessment, then a necessary first step in science instruction is defining the learning outcomes that we are trying to achieve. According to the *National Science Education Standards* (NSES) (National Research Council 1996), goals for science learning can be categorized into the following areas: science subject matter; science inquiry skills and understandings about inquiry; science and society issues; and the history and nature of science. Schools and teachers add their own goals for student learning that go beyond science content, for example, the development of critical-thinking skills, social skills, positive attitudes, and ethical behavior. Our focus here is on assessing student learning of science subject matter, although some examples will provide information on other aspects of student learning as well.

Even if we agree about the goals for science learning, what counts as evidence of learning is in the eye of the beholder. For some, above-average achievement on state-sponsored standardized tests is an indicator of student learning of science subject matter. For others, good performance on textbook end-of-chapter tests means that students have learned. Others count what students say in class or write in laboratory notebooks as evidence of science learning. How can we make reasonable decisions about what counts as science learning?

Recent research on how people learn can help us make such decisions (Bransford, Brown, and Cocking 1999). Bransford and his colleagues pointed out several ways in which students (novices) differ in their learning from experts in the field:

1. "Experts notice features and meaningful patterns of information that are not noticed by novices" (19). Expert chess players, for example, are able to "chunk" information about chessboard configurations that novices do not notice or find meaningful.

2. "Experts have acquired a great deal of content knowledge that is organized in ways that reflect a deep understanding of their subject matter" (19). For example, a physicist will classify problems on the basis of the physics principles to be applied (e.g., momentum), while a student will classify problems based on surface features (e.g., inclined planes). Novices struggle with not only understanding new content but organizing it in meaningful ways.

3. "Experts' knowledge cannot be reduced to sets of isolated facts or propositions but, instead, reflects contexts of applicability: that is, the knowledge is 'conditionalized' on a set of circumstances" (19). Experts know when to apply their knowledge, given the problem-solving situation. For example, expert fly fishermen know which flies to use under which conditions—season, weather, location, time of day, and so on.

4. "Experts are able to flexibly retrieve important aspects of their knowledge with little attentional effort" (19). Expert drivers do not need to think about each step of the process the way beginners do. Their driving is a fluent and automatic response, and they can attend to other activities (like carrying on a conversation—we hope not on a cell phone!) while driving.

5. "Experts have varying levels of flexibility in their approach to new situations" (19). "Artisan" experts are able to perform relatively routinized tasks fluently, while "virtuoso" experts display "adaptive expertise" and respond creatively to new situations. This adaptive process requires the expert to monitor what she understands and what she needs to learn, in other words, to be metacognitive.

From each of these principles of expertise, we can derive expectations for performance that would count as science learning. For example, because experts notice meaningful patterns, our assessments ought to require students to describe patterns in data, or distinguish meaningful similarities and differences among objects, not merely recite definitions. Because experts are able to organize their knowledge in meaningful ways, assessments of learning should ask students to demonstrate their understanding of the relationships among a set of principles. We could consider student success on such types of tasks as evidence of learning. In Table 1-1, we list the five principles of expertise and implications for what counts as evidence of learning. We also present examples of assessment items that would provide such evidence. Thus, our conception of what counts as science learning expands when we consider what the research says about expert learning. However, we do not expect our elementary and middle-level students to become experts in science. According to Berliner (1994), it is more reasonable to expect that students develop over time through closer and closer approximations of competency, proficiency, and eventually, expertise.

Learning science, according to Bransford, Brown, and Cocking (1999), also means that students can transfer what they know to other situations. Transfer is influenced by a number of factors: (1) initial understanding; (2) time; (3) motivation for learning; (4) degree of contextualization or abstraction; (5) monitoring and feedback (metacognition); and (6) previous knowledge, experience, and culture. These factors have implications for seamless assessment. For example, if transfer is influenced by previous knowledge, then assessment must include opportunities to diagnose incoming science ideas. If transfer is influenced by the degree of contextualization or abstraction, then assessments must ask students to apply their learning to situations that are quite similar to instruction as well as those that are new. If transfer is influenced by metacognition, then assessment must include opportunities for students to reflect on what they know and do not know. Using these ideas about learning, we see that memory and reproduction of information are insufficient as evidence of learning. We must also assess the degree to which student learning can transfer to new situations. Table 1-2 summarizes factors that influence transfer and the concomitant implications for assessment.

Seamless assessment is based on a view of learning science in which students understand science concepts, can organize and retrieve them in meaningful ways, and can transfer their understanding to new contexts to solve new problems. To learn about these dimensions of our students' learning, we will need to apply a variety of assessment strategies throughout our science instruction. We anticipate that our assessments will also allow students to learn science more deeply.

Table 1-1 The Implications of Expert Learning for Assessment in Science

Principles of Expertise*	What Counts as Evidence for Learning	Sample Assessment Item
1. Experts notice patterns.	Students' ability to classify, categorize, identify similarities and differences, and see patterns.	Using a set of leaves, make a dichotomous key that works to distinguish each type from the others.
2. Experts organize their knowledge.	Students' ability to structure their knowledge and see relationships among concepts.	Make a concept map that uses the following terms: *plant, animal, producer, consumer, predator,* and *prey.* Include ten more terms of your choice. Label the lines between each concept with a relation word (e.g., *is, for example, includes*).
3. Experts understand ideas within a context.	Students' ability to use a concept to solve problems in various contexts.	Use what you know about pendulums to fix the grandfather clock in the hallway that is five minutes too fast. Explain what you would do and why.
4. Experts flexibly retrieve knowledge.	Students' ability to retrieve the relevant information to solve a problem.	What time will the full moon rise in New York? In Missouri? How did you figure that out? What does the third-quarter moon look like in California? In Australia?
5. Experts approach new problems with varying degrees of flexibility.	Students' ability to recognize what they know and need to know in order to solve a problem.	How can we determine the quality of the stream? What do we know from our data? What more do we to know? How can we find out?

*From Bransford, Brown, and Cocking 1999

Table 1-2 Factors That Influence Transfer of Learning and Their Implications for Assessment in Science

Factors That Influence Transfer*	Implications for Science Assessment
Initial understanding: The learner must understand the concept in order to transfer learning.	Science assessment must ask questions that demand understanding, not merely memorization.
Time: Learning with understanding takes time. Time on task is more valuable when students see the transfer implications of what they are learning.	Science assessments must include opportunities for application to real-life situations.
Motivation for learning: Students will spend more time on learning if they are motivated (by challenge, by relevance, and by social opportunities).	Science assessments must include challenges at an appropriate level, opportunities for social interactions, and relevant contexts.
Degree of contextualization or abstraction: Transfer occurs more readily when concepts are taught using multiple contexts.	Science assessments must use contexts similar to instruction as well as those that are new.
Monitoring and feedback: Transfer is improved when students monitor their learning.	Science assessments must include opportunities for students to be metacognitive about what they know and what they need to know.
Previous knowledge, experience, and culture: What students bring to the learning situation can facilitate or interfere with learning.	Science assessments must diagnose students' incoming ideas and then monitor the change in their ideas throughout instruction.

*From Bransford, Brown, and Cocking 1999

■ Assessment as a Process: What the Standards Say

So far we have seen that classroom assessment can serve a number of purposes and take different forms based on our views of learning. Assessment is a process, not an activity. Once teachers have determined their learning goals, they design instruction and assessment aligned with those goals. Next, they collect assessment information from their students during the course of

Table 1-3 The *National Science Education Assessment Standards* **and Seamless Assessment**

Assessment Standard*	Indicators* Relevant to Seamless Assessment
A. Assessments must be consistent with the decisions they are designed to inform.	• Assessments are deliberately designed. • Assessments have explicitly stated purposes.
B. Achievement and opportunity to learn science must be assessed.	• Collected achievement data focus on the science content that is most important for students to learn.
C. The technical quality of the data collected is well matched to the decisions and actions taken on the basis of their interpretation.	• Assessment tasks are authentic. • Students have adequate opportunity to demonstrate their achievements.
D. Assessment practices must be fair.	• Assessment tasks must be set in a variety of contexts, be engaging to students with different interests and experiences, and must not assume the perspective or experiences of a particular gender, racial, or ethnic group.
E. The inferences made from assessments about student achievement and opportunity to learn must be sound.	• When making inferences about student achievement and opportunities to learn science, explicit reference needs to be made to the assumptions on which the inferences are based.

*From the National Research Council 1996

instruction. Finally, they interpret the evidence and use it in some way. For example, assessment informs teachers about what to do next in their instruction. In addition, assessment information provides feedback to the student and to parents. Sometimes classroom assessment data are used to judge teacher and school quality. This book is most concerned with how teachers can use classroom assessment to inform instruction and help students learn science.

The *National Science Education Standards* (*NSES*) offer guidance about what constitutes quality assessment in science. In Table 1-3, we list the *NSES* assessment standards that are most directly tied to seamless assessment. According to the *NSES*, teachers use classroom assessments for a number of different purposes: to improve classroom practice; to plan for instruction (by determining students' initial understanding and by monitoring student progress); to help students know what is important to learn; to help learners become self-directed; and to report student progress to a variety of stakeholders. Our ideas about seamless assessment are consistent with these standards. Seamless assessment is a vehicle for achieving the standards.

■ Useful Terms

This book is not meant to be a comprehensive textbook on assessment, but rather a users' manual for a particular brand of assessment. However, some technical terms will be helpful to keep in mind as we proceed. Thus, we present a brief list of assessment terms and their definitions.

● *Formative* and *summative assessment* are terms that refer to the timing and the purpose of assessment.

Formative assessment: Formative assessment occurs during instruction. Its purpose is to provide feedback to teachers (and students) as students engage in the learning process. This feedback helps teachers think about the kinds of instructional interventions that need to occur. Formative assessment also helps students monitor their own learning.

Summative assessment: Summative assessment provides documentation of student learning at particular points in time, most typically at the end of a unit of study. Summative assessment information is often the basis of reports to students, parents, and school administrators.

● When scoring assessments, we use *norm-referenced, criterion-referenced,* and *child-referenced* approaches.

Norm-referenced: When assessment data for a given student are compared against a group of other students for the purpose of a score or a grade, that is considered norm-referencing. A test that is graded on a curve is an example of a norm-referenced assessment.

Criterion-referenced: When assessment data for a given student are compared with a performance standard, then the assessment is criterion-referenced. Scoring rubrics provide a method of criterion-referenced assessment.

Child-referenced: When we compare a student's performance at a point in time with her previous performance, we are using a child-referenced approach. Portfolio assessment is an example of child-referenced assessment.

- *Validity* and *reliability* are measures of the quality of an assessment item or task.

  Valid: Assessments are considered valid when they measure what they are supposed to measure.

  Reliable: Assessments are considered reliable if they are accurate measures that provide similar results with repeated administrations.

- *Embedded* and *authentic* are characteristics of some classroom assessments.

  Embedded: Assessment that occurs within and is often indistinguishable from instruction is embedded.

  Authentic: Authentic assessment simulates the kinds of real-world activities of experts.

- Assessments can be either *formal* or *informal*.

  Formal assessments are typically announced to students as such, are scored, and are graded. Tests and final presentations are examples of formal assessments.

  Informal assessments are often embedded in instruction and sometimes are not scored or graded.

Seamless assessment is formative and summative, can be formal or informal, and takes place throughout a unit of instruction. It can be norm-, criterion-, or child-referenced, and it should adhere to standards of validity and reliability. Seamless assessment is usually both embedded and authentic. Yet teachers do not need to be psychometricians to design effective classroom assessment. Instead, they need to keep in mind what is being assessed and why. They need to interpret the evidence from assessments thoughtfully. They then must make reasonable instructional decisions based on their interpretations, including whether to present further examples, engage students in further discussion, provide more firsthand experiences, or reach closure on the topic. Through systematic collection and interpretation of assessment data, teachers become effective at monitoring and improving student learning.

■ Why Seamless Assessment?

Most readers who have followed us to this point will likely agree with our premise that effective assessment of science learning must be part of the ongoing instructional sequence. However, we feel a responsibility to offer a few arguments to justify our approach. Over the past twenty years, the research literature has demonstrated that students come to science class with strongly held ideas about the world that can differ greatly from the accepted scientific explanations (Driver et al. 1994; Wandersee, Mintzes, and Novak 1994). These ideas have been called misconceptions, alternative conceptions, and children's science, to name a few. Once we recognize that our students are not blank slates, but filled with ideas about how the world works, we become responsible for helping them grow in their conceptual understanding. This responsibility necessarily involves assessing their progress as they build their ideas. Seamless assessment is an essential component of any science curriculum that makes student understanding central, because it helps reveal students' ideas and implies next steps in instruction.

Secondly, we live in an information age, where scientific knowledge grows exponentially over shorter and shorter time periods. (A recent report from the National Science Foundation stated that more than ninety-two thousand scientific articles were published in 2001, in comparison with about seventy thousand in 1991 [Hill 2004].) Our science textbooks reflect this knowledge explosion, growing thicker with each new edition. Expecting students to memorize more and more science facts is no longer a reasonable strategy. Helping students understand a few ideas deeply, and facilitating the development of lifelong learning, is a more viable approach. Seamless assessment goes beyond measuring the accumulation of knowledge. It helps teachers teach for understanding and monitor meaningful learning over time.

Finally, we also live in an age of accountability. Teachers are under pressure from administrators, community members, and policy makers to demonstrate high student performance on standardized tests. Yet those test scores cannot begin to capture the complexity of student thinking about living things, the physical world, or earth and space. Teachers must not rely solely on external measures to judge the quality of their teaching and student learning. Teachers must find ways to tell the stories of the learning that is taking place each day in their classrooms. Seamless assessment provides the data for those stories.

Seamless Assessment and the 5 Es

2

A ssessment and instruction are logically linked. We set goals for learning that we translate into instruction from which we hope students will learn. We assess the degree to which students have learned the goals we defined. Yet assessment can be much more; assessment can provide incoming diagnostic information about students' science knowledge. Assessment can let us know what students understand or do not understand so that we can refine our instruction. Assessment can help students learn from each other and from the assessment itself. Assessment can help us determine what students learned. And, assessment can help students reflect on their own learning. Seamless assessment is aimed at accomplishing each of these assessment purposes across a unit of instruction.

Our model of seamless assessment uses current views of inquiry-based instruction as a starting point. Classrooms in which inquiry occurs are characterized by five essential features (National Research Council 2000). According to the National Research Council, inquiry occurs when

- learners are engaged by scientifically oriented questions

- learners give priority to evidence, which allows them to develop and evaluate explanations that address scientifically oriented questions

- learners formulate explanations from evidence to address scientifically oriented questions

- learners evaluate their explanations in light of alternative explanations, particularly those reflecting scientific understanding

- learners communicate and justify their proposed explanations (28)

One instructional model that incorporates these features of inquiry is the 5E model.

■ The 5E Model

We have several reasons for choosing the 5E model (Bybee 1997; 2002; National Academy of Sciences 1998) to guide our science instruction and seamless assessment. First, the 5E model is a useful tool for designing science lessons. It helps us focus on important concepts that we want students to learn and helps us think about an appropriate series of learning opportunities that will best help them learn those ideas. Second, the 5E sequence is based on what we know about how people learn (Bransford, Brown, and Cocking 1999). The 5E model recognizes that learners learn best when they are engaged in doing and thinking, when they have a chance to build new ideas after exploration, and when they can apply their learning to familiar and new contexts. Third, the 5E model directly addresses the research about student misconceptions in science (Driver et al. 1994; Wandersee, Mintzes, and Novak 1994). In a 5E unit, teachers (1) help students make their science conceptions explicit, (2) challenge those conceptions with new evidence, and (3) facilitate student building of scientifically accurate concepts. These strategies have the potential to create deep and lasting understanding in students. Last, but certainly not least, the 5E model is consistent with current views of inquiry. Throughout the five phases of the instructional sequence, students engage in science questions, collect and use data to formulate explanations, and evaluate and communicate their explanations. In other words, the 5E model captures the essential features of inquiry.

The instructional model is composed of five phases: *Engage, Explore, Explain, Elaborate,* and *Evaluate.* Each phase aims at a slightly different purpose for science learning (see Table 2-1). In the *Engage* phase, students encounter a scientific question, idea, or natural phenomenon. Teachers might introduce a unit with a field trip, demonstration, or discrepant event, a problem to solve, a current event, a local issue, a discussion, or some other strategy to engage students' attention and get them thinking about the scientific questions they will encounter in the unit. During the *Explore* phase, students have firsthand experience with a phenomenon. They might carry out investigations using laboratory equipment, make

Table 2-1 The 5E Model of Science Instruction

Model Phase	Learning Purposes*
Engage	• Initiates the learning task. • Introduces the major ideas of science in problem situations. • Makes connections between past and present learning experiences. • Focuses student thinking on the learning outcomes of the upcoming activities. • Mentally engages students in the concept to be explored. • Motivates students.
Explore	• Provides opportunities for students to test their ideas against new experiences. • Provides opportunities for students to compare their ideas with the ideas of their peers and the teacher. • Provides a common base of experiences in which students actively explore their environment or manipulate materials.
Explain	• Provides opportunities for students to develop explanations. • Introduces formal language, scientific terms, and content information to make students' previous experiences easier to describe and explain.
Elaborate	• Applies or extends students' developing concepts in new contexts. • Provides opportunities for students to develop deeper and broader understanding.
Evaluate	• Encourages students to assess their understanding as they apply what they know to solve problems.

*From Bybee 2002; National Academy of Sciences 1998

observations in nature, or collect data using the Internet. Often students work in teams to build a common base of experience about the phenomenon under study. The purpose of the *Explain* phase is for students to formalize their understanding of the concepts under investigation. Students invent explanations and use evidence from the *Explore* phase to support their ideas. Teachers introduce formal ways to represent these ideas—terms, formulas, diagrams, and so on. In the *Elaborate* phase, students build on their understanding by solving new problems in new contexts. Teachers design ways for students to extend what they know by transferring their understanding to these problems. The *Evaluate* phase provides opportunities for students to reflect on and demonstrate what they know. Students communicate their learning to various audiences.

■ Five Es and an A

Each phase of the 5E model has implications for assessment. For example, during the *Engage* phase, if instruction is to be effective, teachers must identify what students already know about the phenomenon under study. This will include understanding commonly held misconceptions and diagnosing their own students' views. In the *Explore* phase, teachers will use formative assessment to understand the progress that students are making in understanding concepts. Finding out what students do not yet understand helps guide the design of new instructional interventions. During the *Explain* phase, teachers listen carefully to student explanations and determine what ideas need further instructional attention. In the *Elaborate* phase, teachers see how well students can use their new understandings and formal representations to solve problems. During the *Evaluate* phase, there is an opportunity for summative evaluation. In addition, in the *Evaluate* phase, teachers can help students be metacognitive about what they have learned and what they still need help in understanding. In Table 2-2, we link each 5E phase with a particular assessment purpose.

These different assessment purposes can be connected with a set of concomitant assessment strategies. For example, diagnosing students' incoming ideas during the *Engage* phase can be accomplished through discussion, by using a KWL chart, by asking students to draw a concept map, or through interviews with selected students. Formative evaluation during the *Explore* phase might take the form of student science notebooks or a demo memo. Summative evaluation in the *Evaluate* phase might include a postconcept map, constructed-response test items, or a presentation. Oftentimes the same activities that are planned to carry out instruction within a phase can be used to carry out assessment. When this happens, assessment has become seamless.

■ Assessment Strategies for Each Phase

The rest of this chapter highlights commonly used assessment strategies, linking them to the 5E phases and assessment purposes displayed in Table 2-2. We present a description and a practical example of each strategy and discuss which vignettes (from Chapters 3 through 5), if any, demonstrate the strategy. The Appendix provides a list of these same assessment strategies with accompanying websites where more detailed descriptions and additional examples can be found. Although we list the strategies by 5E phase, they are meant to be used flexibly. Some can be used in a variety of 5E phases for

Table 2-2 Assessment Purposes and the 5E Model

Model Phase	Assessment Purposes
Engage	• For teachers to identify students' incoming science ideas and misconceptions. • Helps teachers determine what students need to explore in the next phase.
Explore	• For teachers to determine how students are progressing in their conceptual understanding. • For teachers to understand what students do not understand and determine instructional interventions that need to occur. • Helps teachers determine what needs to be explained in the next phase.
Explain	• For students to demonstrate their current understanding. • For teachers to determine what ideas need further instructional attention. • Helps teachers determine what elaborations will help scaffold learning in the next phase.
Elaborate	• For students to demonstrate their ability to apply and transfer their understanding to new contexts. • For teachers to see how students use formal representations of science knowledge (terms, formulas, diagrams). • Helps teachers determine what will be important to evaluate in the next phase.
Evaluate	• For teachers to determine what students learned from the unit. • For students to be metacognitive about their learning. • Helps teachers make decisions about new 5E learning cycles.

various assessment purposes. For example, notebooking, predicting, and concept mapping are versatile forms of assessment that can occur throughout the 5E cycle. The following list includes assessment strategies that are illustrated in the vignettes as well as some that are not. Furthermore, the vignettes include other assessment strategies not in the following list. The assessment strategies listed below, together with Tables 2-1 and 2-2 and the vignettes, constitute a toolbox for designing seamless assessment in science. The committed reader will be creating new assessment strategies to add to this list in no time!

Assessment Strategies for the Engage Phase

KWL chart (Ogle 1986). This assessment strategy helps students think about what they **K**now about a topic, what they **W**ant to know about it, and after they have finished, what they **L**earned. The teacher begins by asking students to complete the first two columns of the KWL chart either in small groups or as a whole class. The list of responses identifies students' existing science ideas and helps the teacher plan instruction. For example, students list that they know the moon is visible only at night. In response, the teacher plans daytime observations of the moon during the *Explore* phase. As the unit progresses, evidence-based ideas are added to the third column of the KWL chart. See "Toiling in the Soil" in Chapter 5 for an example of how this strategy can be used during the *Engage* and *Elaborate* phases.

Concept mapping (Novak 1998; Novak and Gowin 1984). Concept maps help students develop a visual representation of their network of knowledge. A network consists of nodes and links. Nodes represent concepts and links represent the relations between concepts. Teachers often begin the strategy by giving each student a topic and a list of nodal terms. Students arrange the terms into connected nodes and label the connecting links. For example, in a unit on water quality, the teacher might provide terms such as *temperature*, *turbidity*, *conductivity*, *tolerance*, and *macroinvertebrates* and ask students to connect the terms. If a student's map indicates that he thinks that water temperature is not related to turbidity, then the teacher can plan explorations of the relationship. This assessment strategy is useful in the analysis of student learning at a variety of 5E phases, and it is often used in a pre/post manner. See "It's Volcanic!" in Chapter 5 for an example of how this strategy can be used during the *Engage* and *Evaluate* phases.

BOX 2.1. ASSESSMENT STRATEGIES FOR THE *ENGAGE* PHASE

KWL chart
concept mapping
card sort task
memoir
brainstorming
interview
questionnaire
Venn diagram
science notebook
predicting
scientists' meeting
observation

Card sort task (Friedrichsen and Dana 2002). Card sorts help students distinguish between two or more potentially confusing ideas. Students examine a number of examples written or drawn on cards. They sort the cards and write a description of the properties of each sorted pile. For example, students often confuse the ideas of melting and dissolving. The teacher can supply cards with pictures of melting and dissolving for students to sort. When she inspects a student's pile and finds cards misplaced, she can use this information to plan an exploration of the phenomena of melting and dissolving. See "Seeds and Eggs" in Chapter 3 and "May the Force Be with You" in Chapter 4 for variations on this strategy.

Memoir (George 2005). This strategy prompts recall of prior learning experiences and demonstrates to students what they currently know about a topic. Thus, it is useful for stimulating prior knowledge at the beginning of a unit or prompting metacognition near the end of a unit. For memoir writing, teachers ask students to think about a topic, what they used to know, and what they now know. For example, at the beginning of a unit a teacher might ask students to think about their magnet history, make claims about magnet behavior, and describe where they learned this information about magnets. If a student claims that magnets can attract hair, because she has noticed the way her hand attracts fur when she pets the family cat, then the teacher might plan an exploration of magnetism and static charge. See "It's Volcanic!" in Chapter 5 for an example of how a memoir can be used as an *Elaborate* phase activity.

Brainstorming. This strategy focuses students on a problem and asks them to generate multiple solutions. Ideas are developed as fast as possible, and judgment is suspended. For example, students observe a burning candle and generate as many properties as they can in two minutes. The teacher does not evaluate the responses, but he encourages students to build on each other's responses. If some of the students say that the wick alone is burning and the wax is simply melting, the teacher might use this information to plan an exploration of the role of wax in burning. See "It's Volcanic!" in Chapter 5 for an example of brainstorming in connection with concept mapping.

Interview or questionnaire (Osborne and Freyberg 1985). Face-to-face interviews or written questionnaires about instances or events probe students' understanding of a specific concept. For example, in an interview or questionnaire about living things, a kindergarten teacher might ask students which things are living things (e.g., person, animal, tree, fire, bicycle). In an interview or questionnaire about the event of light and vision, a sixth-grade teacher might ask students to explain how light helps a person see a tree. Their responses can be used to plan follow-up explorations.

"Misconceiving the Moon" in Chapter 5 uses a questionnaire about moon phases to begin the unit.

Venn diagram. This strategy exposes how students organize their knowledge. Large circles represent properties of different groups, with properties shared across groups at the intersection. For example, students might use a Venn diagram to represent the similarities and differences among mammals, reptiles, and birds. The teacher may use this information to plan lessons about animal groups or evaluate how student understanding has progressed. See "What's the Anther?" in Chapter 3 and "May the Force Be with You" in Chapter 4 for examples of how Venn diagrams can be used at various 5E phases.

Assessment Strategies for the Explore Phase

Science notebook (Campbell and Fulton 2003). The science notebook reveals how students respond to field trips, laboratory activities, and problems throughout a unit of study. The science notebook has many uses, including keeping an accurate set of observations, recording quantitative data, describing the context or procedures for a lab or field experience, writing and drawing ideas to clarify thinking, making connections, wondering, and contemplating. For example, in response to a field trip to a canyon, a student may write: "Today, I was climbing in Big Cottonwood Canyon. The rock there is pinkish. The geologist said it was quartzite. What is quartzite? It feels smooth like limestone. There are white veins in it and it is very blocky. The edges are almost square." This assessment strategy gives ongoing information about how students are building conceptual understanding and enables the teacher to assess progress. See "What's the Anther?" and "Water You Know" in Chapter 3, "Shocking News: Static Electricity" in Chapter 4, and "Toiling in the Soil," "Rock On!" and "Misconceiving the Moon" in Chapter 5 for examples of how science notebooks can be used.

BOX 2.2. ASSESSMENT STRATEGIES FOR THE *EXPLORE* PHASE

science notebook
conceptual cartoon
think, pair, share
drawing completion
predicting
demo memo
scientists' meeting
chart
observation

Conceptual cartoon. This strategy applies a science concept to a real-world situation depicted by a cartoon drawing and forces students to make a choice. The conceptual cartoon is designed to intrigue, to provoke discussion, and to stimulate scientific thinking. For example, a teacher might present a cartoon of a snowman with several students standing around it with cartoon bubbles above their heads. One is saying, "Don't put the coat on the snowman, it will melt him." Another is saying, "Leave the coat on, it will help the snow stay cold." And a third is saying, "I think it doesn't make any difference." The teacher presents the cartoon and the students discuss the various claims and provide evidence for their choices. The discussion provides information about the students' understanding of heat transfer and helps the teacher decide what kinds of investigations might help students solve the conceptual cartoon.

Think, pair, share (Victor and Kellough 2000). This strategy provides a quick check of student thinking about the concepts being taught. The teacher poses a challenging question and gives students a short time to think about the question individually; then they discuss their thoughts in pairs. Finally students share their answers with the class. For example, the teacher might ask, "Do you think a seed is alive? Why do you think that?" This strategy samples ongoing student understanding as part of a discussion and tells the teacher what needs further exploration and explanation. See "What's the Anther?" in Chapter 3 for an example of how this strategy can be used.

Drawing completion. Drawing helps students visualize and develop understanding of a phenomenon. The teacher poses a challenge and may provide a partial diagram. Students draw or complete the diagram and use the drawing to help solve a challenge. For example, the teacher may ask, "What is the smallest mirror you can use to see your entire body?" He gives students a drawing of a boy standing in front of a wall and asks them to draw a mirror on the wall and the lines of sight the boy would use to see his entire body. Student responses help the teacher decide what needs further exploration or explanation. See "What's the Anther?" in Chapter 3, "Mirror, Mirror on the Wall" in Chapter 4, and "Toiling in the Soil" and "It's Volcanic!" in Chapter 5 for examples of how this strategy can be used.

Predicting activities. This strategy requires students to make evidence-based predictions about a demonstration or a hypothetical situation. The instructor provides a context and asks students to make a prediction about what will happen next or what will happen if some variable is changed. For example, after introducing pendulums with a story of Tarzan swinging on vines, the teacher asks, "What will happen to the swing of the vine if Jane joins Tarzan on the vine?" Student responses influence next steps. See "Mirror, Mirror on the Wall" and "Will It Float?" in Chapter 4 for examples of how this strategy can be used.

Demo memo. Demonstrations can help all students witness and discuss the same phenomenon. The demo memo requires students to write a brief summary of the essential features of a student- or teacher-conducted demonstration. After a demonstration is completed, students describe what happened and how the demo provided an example of the underlying principle. For example, after demonstrating pitch by using bottles with different amounts of liquids and both blowing into the bottles and tapping them with a metal spoon, the teacher asks students to write about what happened and why they think the pitch was different in the two cases. See "Mirror, Mirror on the Wall" in Chapter 4 for another example of how this strategy can be used.

Assessment Strategies for the Explain *Phase*

Exit sheet; exit ticket; minute paper. These assessment strategies reveal (to teachers and to students) what is clear and unclear about a given concept. The teacher asks students to write brief responses to a few questions at the close of the class period (e.g., What is one thing you learned? What is one example of _____? What is one thing that was unclear? Can you explain how _____?). The teacher collects all responses and reads them prior to the next class. For example, the teacher might ask, "What is something you learned about sinking and floating and something that was unclear about sinking and floating?" Student responses provide information about possible instruction needed in subsequent classes. See "It All Goes Back to Plants" in Chapter 3 and "May the Force Be with You" in Chapter 4 for examples of this strategy in use.

BOX 2.3. ASSESSMENT STRATEGIES FOR THE *EXPLAIN* PHASE

exit sheet; exit ticket; minute paper
discrepant event
ConcepTest
making a model
making a claim; theory choice
meaningful paragraph
science notebook
KWL chart
labeled drawing
predicting
letter to the teacher

Discrepant event (Liem 1987). The discrepant event is a demonstration or activity with an unexpected outcome. It challenges students to use newly formed concepts to revise their explanations to account for this outcome. The teacher asks students to predict what they think will happen, to observe what actually happens, to revise their explanations, and to think about possible tests of their new explanations. The sequence of prediction, observation, and explanation is critical. In a study of density, many students think that big things sink and little things float. The teacher brings out a massive log from an oak tree and a huge tub of water and asks the students to predict if the log will sink or float. The surprise the students experience when they witness the log floating causes them to reevaluate their theories of sinking and floating. Student responses help the teacher select an appropriate elaboration activity. See "Will It Float?" in Chapter 4 for another example of how discrepant events can be used in seamless assessment.

ConcepTest (Mazur 1997). The ConcepTest consists of well-posed multiple-choice questions positioned at critical points in a unit and presented aloud to the entire class. These questions probe depth of understanding of important science concepts without invoking equations or formulas. A ConcepTest question focuses on a single concept and provides obvious and counterintuitive foils. Once students have made their individual choices, the teacher asks for a show of hands supporting each foil. Pairs of students then discuss the reasons for their choices. The teacher asks for a second show of hands, which often results in more correct selections. The teacher then provides the correct choice and asks students to write their explanation for why that choice is correct. For example, the teacher poses the following buoyancy question to the class: "A boat carrying a large boulder is floating on a lake. The boulder is thrown overboard and sinks to the bottom of the lake. Will the level of the water in the lake with respect to the shore (a) go up, (b) go down, or (c) stay the same?" (correct choice is b). Depending on how students respond, the teacher will decide how well students understand buoyancy and what elaboration activity might help them increase the depth of their understanding.

Making a model. This strategy uses three-dimensional models to help students visualize and explain aspects of the natural world that are otherwise impossible to see or difficult to imagine. The teacher can assign small groups of students to design a three-dimensional model of the phenomenon or can present her own model. Students discuss the models and provide explanations for how a model helps explain evidence collected in the *Explore* phase. The teacher can then quiz students on aspects of the model. For example, the teacher may assign students to make a scale model of our solar system, using analogies to represent relative planetary size. One student group uses musical notes for their model, representing planetary size by the length of the note (e.g., sixteenth notes versus whole notes). The accuracy of the model is a good way to assess student understanding and to select elaboration activities that build on what they know. See "Misconceiving the Moon" in Chapter 5 for another example of how models can be used.

Making a claim. This strategy challenges students to make evidence-based claims about how something will behave in accordance with concepts they have been learning. The students work together in teams to construct their claims. The class evaluates each claim based on the concept under study and the evidence that was gathered during the *Explore* phase. For example, after investigating sinking and floating, students write claims on overhead transparencies. One group writes, "Objects with a density greater than one will sink." The teacher invites students to the front to present their claim and invites the class to agree or disagree based on evidence from earlier investigations. The quality of the claim and of the audience critique helps the teacher decide next steps. See "Sounding Off" in Chapter 4 and "It's Volcanic!" in Chapter 5 for other examples of how this strategy can be used.

Meaningful paragraph. This strategy is used at the close of the *Explain* phase to assess the depth of students' understanding of terms that are related to a central concept. The teacher provides these terms and students use them to construct a meaningful paragraph. For example, the teacher may assign the students to write a paragraph about the water cycle using the following terms: *rain*, *lake*, *evaporate*, *condense*, *cool*, and *heat*. Students use the last few minutes of class to create their paragraphs and then hand them in. The teacher assesses each reply with a ✓+, ✓, or ✓–, depending on accuracy. Ideally, students learn how well they understand the material and the teacher finds out how well students can communicate the technical terms and the relations among concepts.

Assessment Strategies for the Elaborate *Phase*

Application problem. Application problems exercise analytic reasoning skills as students make connections between a new phenomenon and a recently learned concept. The teacher introduces the phenomenon to small groups of students. The students use the concepts they have learned to develop an explanation. The students either present verbal explanations (to the class) or written ones (to the teacher). For example, after studying the motion of pendulums, the teacher asks students to solve the following problem: "How many ways can you come up with to fix the swing in my backyard, which swings crookedly?"

```
┌─────────────────────────────────────────────────────────┐
│  BOX 2.4. ASSESSMENT STRATEGIES FOR THE ELABORATE PHASE   │
│                                                           │
│   application problem                                     │
│   pair problem solving                                    │
│   puzzler                                                 │
│   thought experiment                                      │
│   debate                                                  │
│   writing and analyzing fiction                           │
│   design activity                                         │
│   science notebook                                        │
│   identification game                                     │
│   team report                                             │
│   data table and graph                                    │
│   predicting                                              │
│                                                           │
└─────────────────────────────────────────────────────────┘
```

The students think about the question and try to solve it using what they know about center of mass. Their solutions provide evidence that they do or do not understand the concepts well enough to apply them to real-world phenomena. Application problems are used as assessment strategies in "Seeds and Eggs" and "It All Goes Back to Plants" in Chapter 3, "Sounding Off" and "Mirror, Mirror on the Wall" in Chapter 4, and "Toiling in the Soil" in Chapter 5.

Pair problem solving (Pestel 1993). This strategy involves pairs of students communicating how to solve a problem. The teacher poses a problem and provides time to solve it. The two students take on the specific roles of problem solver and listener. The problem solver reads the problem aloud and talks through the solution. The listener follows along and asks questions if the problem solver's thought processes are unclear. Students switch roles for each problem. Presenting one's problem solution aloud to a partner helps make reasoning skills explicit. For example, students are given two electricity problems. The first problem asks which set of batteries will last longer—the set powering one bulb or the set powering two bulbs in series. The second problem asks which set of batteries will last longer—the set powering one bulb or the set powering two bulbs in parallel. The resulting discussion helps students think aloud and to listen to how others reason and helps the teacher understand the quality of student thinking.

Puzzler. Puzzlers (Abell, George, and Martini 2002) tease students with a beguiling problem about the topic they are studying. The teacher poses a question and students individually write an answer. Time may be provided for actual experimentation if needed. A few students might be selected to read their answers out loud. Finally, the teacher leads students in a whole-class discussion. For example, during a levers unit, the teacher may pose this question at the start of class: "How could I weigh a heavy object if I did not have a scale that measured that much?" Student responses help the teacher decide what, if anything, needs further elaboration. See "Misconceiving the Moon" in Chapter 5 for an in-depth example of how this strategy can be used.

Thought experiment. The thought experiment reflects the history of science and engages student imagination to investigate nature. Galileo and Einstein were, arguably, the most impressive thought experimenters in the history of science. The teacher frames a problem and students work through a logical solution without actual materials. The students work individually or in teams to approach the problem by thinking (and drawing) possible solutions. For example, after studying levers, students are asked to build a contraption for the zoo that will lift the smelly elephant off its favorite blanket to give it a bath. The students use what they know about force and distance in levers to draw a solution and explain how it will work. This activity challenges students to think deeply about and transform science concepts into useful solutions. "Misconceiving the Moon" in Chapter 5 discusses a thought experiment presented to students in the form of a puzzler (see previous strategy).

Debate. Debate engages students in examining two sides of an issue by adopting the role of a stakeholder. The teacher provides a controversial issue, debate guidelines, resources, and time limits. Students review appropriate debate behavior, research various perspectives, gather pertinent evidence, develop a position, and represent that position in an appropriate format (e.g., town meeting). In a variation called *structured controversy* (Johnson and Johnson 1985), students have to prepare and defend both sides of the issue. For example, the teacher may pose this statement: "Hunting is a necessary tool for wildlife management." Students must use what they have learned about predator-prey relations and carrying capacity to effectively debate the issue.

Writing and analyzing fiction. Trying to write stories that incorporate science concepts is a creative way for some students to synthesize their learning. Teachers generate the story-writing context and young authors take off. For example, near the end of a unit about erosion, students might write stories in which a time traveler witnesses the before and after scenarios of various landforms. "Water You Know" in Chapter 3 uses eco-mysteries as a form of seamless assessment. In addition to writing fiction, students can also be asked to analyze works of fiction for the accuracy of the science concepts displayed. For example, near the end of a unit on sinking and floating, the teacher might read aloud the picture book *Who Sank the Boat?* (Allen 1982) and ask students if they think the story is an accurate representation of the science concepts under study. "Misconceiving the Moon" in Chapter 5 provides an example of this

Elaborate phase strategy, while "It's Volcanic!" in Chapter 5 uses literature to *Engage* students in the topic.

Design activity. A design activity is an engineering challenge that requires students to use science concepts to develop new products. Scale models or working prototypes are often included in the project. The teacher provides guidance, deadlines, and an audience for the final demonstration. The students work in teams or individually to design something that addresses the challenge. For example, after learning about force and motion, students design a roller coaster that is artistic, technically feasible, and safe. This activity challenges students to be creative while using their science knowledge. "Shocking News: Static Electricity" and "Will It Float?" in Chapter 4 use design activities as summative assessments in the *Evaluate* phase.

Assessment Strategies for the Evaluate *Phase*

Poster. Asking students to make a poster invites them to tell a story that includes a statement of the problem, a description of the method, a presentation of results, and a summary of the work in an engaging format. The poster is one refreshing alternative to the laboratory report. The teacher provides materials, project guidelines, a due date, and an audience for the final presentation. Students work singly or in groups to present findings. For example, after studying growth and nutrition in plants, students make posters about their investigations on the effects of various chemicals on plant growth. This activity challenges students to think creatively as they integrate their understanding into the poster. See "What's the Anther?" in Chapter 3 for an example of how posters can be used during the *Explore* phase.

BOX 2.5. ASSESSMENT STRATEGIES FOR THE *EVALUATE* PHASE

poster
constructed response
presentation
comparison essay
final reflection
self-evaluation
one-page memo
scenario exam
concept mapping
Venn diagram
science notebook

Constructed response. The constructed-response test item consists of an open-ended, short-answer question that measures application skills as well as content knowledge. The questions often draw upon authentic, real-world examples. The teacher may include constructed-response items as one part of a unit test. Students' answers demonstrate their understanding in greater depth than would multiple-choice responses. The teacher typically generates rubrics to score responses in terms of specific goals of instruction. For example, after studying ecosystems ecology, the teacher asks students to draw and explain a food chain that includes three levels. The constructed response challenges students to recall facts, understand concepts, and report in a logical way. Since many standardized tests use the constructed-response format, this assessment helps prepare students for high-stakes testing. See "Seeds and Eggs" in Chapter 3 and "May the Force Be with You" and "Sounding Off" in Chapter 4 for other examples of constructed-response items.

Presentation. This strategy challenges students to organize a presentation (using PowerPoint or other visual aids) that demonstrates what they have learned. The teacher provides presentation guidelines, tutorials on how to use technology, and a due date. The students use their ingenuity to develop an interesting presentation that demonstrates the most important points of their learning. For example, students might create seven PowerPoint slides demonstrating what they have learned about heat and temperature. See "It All Goes Back to Plants" and "Water You Know" in Chapter 3 for examples of how this strategy can be used.

Comparison essay. This assessment strategy requires students to compare and contrast the similarities and differences between two categories of things. The teacher provides a written description of the assignment and the deadline for completion. Individually or in pairs, students write an essay comparing two things or two aspects of something the class has been studying. For example, after studying reproduction in flowering plants, students are assigned to compare tomato and corn reproductive parts (e.g., stamen, pistil, pollen, and eggs), giving equal attention to the two plants. An example of a class comparison strategy appropriate for younger students, using a comparison chart instead of an essay, is illustrated in "Seeds and Eggs" in Chapter 3.

Final reflection. The final reflection requires students to summarize their current understanding of the topic or concept, using diagrams, verbal explanations, formulas, and/or data from class observations. The teacher provides guidelines of what to include in the final reflection and encourages students to use their science notebooks as resources. Students work individually to write reflections on their learning. For example, after a study of current electricity, the teacher assigns students to use their electricity notebooks to explain how electricity is involved in making bulbs light and how series and parallel circuits are formed. She asks students to include diagrams, written explanations, and data from class activities to substantiate their ideas. See "Misconceiving the Moon" in Chapter 5 for an example of how this strategy can be used.

Self-evaluation. In self-evaluation, students reflect on their work and the depth of their understanding, often comparing their current ideas with their incoming ideas. The teacher provides prompts such as: "Before I thought _____; now I think _____," or "I could have done better if I had _____." Students respond in writing and the teacher remarks on the self-evaluation. The teacher's comments help students develop new learning goals. For example, at the end of a unit about plant reproduction, students complete the self-evaluation by writing one idea they had about plants that changed during the unit and one question they still have. Thus self-evaluation requires students to be metacognitive about their understanding and to make plans to improve their learning. "Misconceiving the Moon" in Chapter 5 illustrates the use of self-evaluation within a final reflection activity (see previous strategy).

One-page memo. The memo provides a more authentic form of assessment than the traditional lab report. The teacher creates a challenge based on the concept under study and provides guidelines for memo writing. Students write a one-page memo to a fictitious boss to report the findings of an investigation. Synthesizing their procedures and findings into a short piece of writing requires students to find the most important ideas in their work to report. For example, after completing a unit about mixtures and solutions, the teacher says, "It is time to compose a report to your boss at Acme Chemicals. Your boss is a busy woman and does not have time to read a ten-page laboratory report. Instead, she requires you to write a one-page memo in which you highlight your research question, procedures, findings, and conclusions and tell how your findings will benefit Acme Chemicals." See "Will It Float?" in Chapter 4 for another example of how this strategy can be used.

■ Seamless Assessment in Action

These assessment strategies, linked to the 5E phases, illustrate the myriad of options available to science teachers who want to integrate assessment more fully into inquiry-based science instruction. In the following chapters, seamless assessment comes alive in the stories of classroom teachers, grades 1–8, who have used many of these strategies to plan instruction and examine student understanding.

3 | Seamless Assessment in Life Science

■ Introduction

The vignettes in this chapter are based on 5E units of instruction carried out with primary (grades 1–3), intermediate (grades 4–5), and middle-level (grades 6–8) science students. In these units, teachers aimed to develop students' conceptual understanding about organisms, life cycles and reproduction, and ecosystems. Each vignette illustrates how teachers used seamless assessment to plan and inform their instruction. Table 3-1 describes the grade levels, topics, and assessment strategies used in the vignettes.

Life science topics are commonly taught in elementary and middle school classrooms. Many teachers enjoy teaching life science topics, perhaps because they have confidence in their own understanding or have a close personal connection to nature. In our classrooms, studying the life sciences should take students beyond knowing the names of plants or the classification of the animal kingdom, beyond planting bean seeds or observing classroom animals. Studying the life sciences should address important big ideas about the living world. That is what the units in this chapter attempt to do.

In the first vignette, Sandra Abell observes a first-grade teacher who asks her students, "How is a plant different from an animal?" Students grow plants with a purpose—to understand how some plants start as seeds, whereas some animals start as eggs. This age-appropriate investigation of a basic biological idea helps students build a foundation for learning other life science concepts.

In the next vignette, Michele Lee works with a group of fourth graders to help them recognize the link between plants and animals. Most of the inquiry that her students engage in is minds-on, rather than hands-on. With the

Table 3-1 Life Science Vignettes by Grade Level, Topic, and Assessment Strategy

Vignette	Grade Level	Topic	Assessment Strategies
Seeds and Eggs	1	Plants versus animals	*Engage*: sorting sheet; scientists' meeting *Explore*: individual science notebook drawings; group alike/different chart; scientists' meeting *Explain*: whole-class plant/animal chart *Elaborate/Evaluate*: group plant/animal identification game; constructed-response answering sheet
It All Goes Back to Plants	4	Importance of plants	*Engage*: group discussion with teacher-directed questions *Explore*: collaborative chart *Explain*: exit ticket *Elaborate*: team report *Evaluate*: scenario exam
What's the Anther?	6	Plant reproduction	*Engage*: field trip notebook; discussion *Explore*: Fast Plant drawings; poster *Explain*: diagram labeling *Elaborate*: think, pair, share; Venn diagram intersection *Evaluate*: cycle diagram; student-generated Venn diagrams
Water You Know	7	Water ecosystem	*Engage*: water meter *Explore*: water quality survey *Explain*: water quality table and class booklet *Elaborate*: eco-mystery *Evaluate*: conference presentation

guidance of an expert teacher, the students use their thinking skills to make sense of an important life science idea that all animals depend on plants for food. Michele's vignette illustrates the care that teachers must take when helping intermediate-age students connect several ideas into a coordinated whole.

In the third vignette, Marsha Tyson and Kelly Turnbough report on their collective experience with teaching reproduction at the middle level. Although the topic is personally relevant for middle-level students, it is often taught as a set of terms to be memorized. Marsha and Kelly use a structure-function approach instead, which engages students in the type of thinking of many biologists. Furthermore, the teachers help students draw comparisons between plant and animal reproduction, returning to some of the basic ideas introduced in the first vignette.

As students mature in their knowledge of life sciences and their reasoning skills, they are ready to tackle more complicated relationships involving both living things (biotic factors) and the nonliving world (abiotic factors). In the final vignette of this section, Sara Torres provides a window into a yearlong investigation of water quality, using the local community as a site for data collection. Sara's story illustrates how teachers can help middle-level students compare their evidence with regional databases and national water quality standards to better understand their local environment.

Each life science unit described in the vignettes was designed and carried out at a particular grade level. However, the *National Science Education Standards*, upon which each unit was built, address grade ranges, not specific grades. We hope you will find ways to adapt these units to your local context and grade level, where age appropriate. We also encourage you to develop new units about other important life science concepts for elementary and middle-level students, as detailed in the standards within the broad categories of life cycles, organisms and their environments, heredity, diversity and adaptation, and regulation and behavior (see National Research Council 1996).

Vignette

Seeds and Eggs

■ *Sandra Abell*

Unit Notes

Grade: 1

Learning Goals: Students will understand that (1) plants grow from seeds and animals grow from eggs and (2) plants and animals have some characteristics in common and differ in other characteristics.

National Science Education Standard: Content Standard: K–4, Life Science: Characteristics of organisms. (See essay, National Research Council 1996, 127–28.)

Assessment Strategies:

 Engage: sorting sheet; scientists' meeting

 Explore: individual science notebook drawings; group alike/different chart; scientists' meeting

 Explain: whole-class plant/animal chart

 Elaborate/Evaluate: group plant/animal identification game; constructed-response answering sheet

Note: An online version of this vignette, complete with classroom video and student artifacts, is available at www.coe.missouri.edu/~abells/roes/class1.html.

Vignette

Engage

"Teacher, these aren't seeds; they're candy," quipped Joel when Mrs. Schwartz came to visit him and his partner at their table. They were engaged in an introductory activity in the "Seeds and Eggs" unit. Their teacher had provided a large piece of paper with two circles, one labeled "seeds" and one "not seeds," and a bag of objects including seeds, BBs, candy, and beads. She asked the students to glue the objects onto one of these circles and to leave out any that they were not sure about. The teams eagerly went about their task while Mrs. Schwartz circulated, encouraging partners to talk to each other, explain their reasoning, and reach consensus if possible. She listened carefully to understand their incoming ideas about seeds.

When the teams had completed this task, Mrs. Schwartz invited them to the scientists' meeting (Reardon 1993), marked by a circle of masking tape on the floor in the front of the room. Students sat on the line and shared their sorting charts. Students agreed on many of the items in the seeds category. When asked why he placed something in the seeds circle, Alex explained, "'Cause I know what all seeds are." Yet there were several items that the students did not agree about, including sliverlike flower seeds and large peas.

"How can we find out if these are seeds or not seeds?" Mrs. Schwartz probed.

"We could look them up in a book."

"We could ask a farmer."

"We could plant them and see if they grow." This was the first of many times that the students would zero in on an important characteristic of living things—growth—during the "Seeds and Eggs" unit. The discussion was a vehicle for assessing student progress and transitioning into the exploration activity.

Explore

The next day Mrs. Schwartz brought in materials the students could use to see if the undecided items in their collection would grow or not. They were to make gardens by "planting" each of those items in a wet paper towel inside a small plastic bag.[1] After the teams completed their planting, each student received an individual science notebook to keep track of his or her observations. The notebook consisted of plain paper folded in half and stapled down the middle with a construction paper cover. Over the next several days students observed their bag gardens and recorded their observations. After a few days, they were ready to sit in the scientists' meeting again and share their results.

"They sprooted!"

"They must have been plants, because they growed."

"We changed our minds. Now we know they are plants."

Mrs. Schwartz decided that the students understood that seeds were things that would sprout and grow. Thus they were ready to compare seeds and plants with another kind of living thing.

Mrs. Schwartz presented what she believed would be a discrepant event to challenge students' thinking about seeds. She gave each team two plastic jars filled with water and organisms. In one jar, labeled "Big Ones," were clover seeds and tap water. In the other jar, labeled "Little Ones," were brine shrimp eggs and saltwater. Without describing the exact contents of the jars, Mrs. Schwartz asked the students to predict what would happen to the Big Ones and the Little Ones. Every group predicted that the Big Ones would sprout, and most thought that the Little Ones would too. Teams observed the jars over time, keeping a record of their observations in team notebooks. After several days of observation, Mrs. Schwartz asked each team to complete a chart about how the Big and Little Ones were alike and different. By this time, the students noticed that the Big Ones (clover seeds) had sprouted, confirming student predictions that the Big Ones were seeds. The Little Ones had not changed much. Nevertheless, in several notebooks Mrs. Schwartz found that students had written, "Both have sprouts," in the "Alike" column. This told her that further observations were needed.

It took a few more days to notice any changes in the Little Ones. During one class period, students observed that the Little Ones were moving, or as Leslie said, "They wiggle and dance around." Together in the scientists' meeting, the students decided that the Little Ones must not be seeds after all. Alicia suggested that they were "eggs."

Mrs. Schwartz posed a question: "If they are eggs, what do you think they will become?" Students generated a flurry of predictions: tadpoles, fish, salamanders, and sea monkeys.

Explain

When the students introduced the term *eggs* during the scientists' meeting, Mrs. Schwartz deemed that they were ready to discuss the big ideas of the "Seeds and Eggs" unit: how plants and animals are alike and different. The following class period, she invited students back to the scientists' meeting, where she had a whiteboard ready with two columns: "Plants" and "Animals." First, she asked students to list claims about how plants and animals are different. Students stated that seeds sprout into plants and eggs become animals. They noted that plants have stems and leaves and roots, while animals have legs and faces. They claimed that animals moved, ate, and breathed, while plants did not. When Mrs. Schwartz asked how plants and animals are the same, the students agreed that both plants and animals grow, from either seeds or eggs. They supplemented their claims with evidence from their lives and from the "Seeds and Eggs" unit. Although these first graders' biological ideas were incomplete in comparison with an expert's, Mrs. Schwartz decided that their claims were age appropriate and reasonable for the purposes of the unit. However, she wanted to see if they could apply their ideas to a new situation.

Elaborate/Evaluate

"NASA has recently found two new life forms in a little-explored part of Indiana. First graders, you have been selected by NASA to be the scientific team to study these life forms and report back to NASA. Is the life form a plant or an animal? Why do you think so?" This application problem was the test that Mrs. Schwartz gave her students on the final day of the "Seeds and Eggs" unit. She distributed two organisms to each team: *Daphnia*, a water flea, and *Anacharis*, a water plant (both commonly available in aquarium supply stores). She encouraged students to use the "Plants/Animals" chart from the previous class to help with their reasons. In addition, she gave each group two new pages for their team notebooks with constructed-response items that read, "We think *Daphnia/Anacharis* is a _____ because _____ ," with plenty of space for writing several reasons. All teams responded that *Daphnia* is an animal and *Anacharis* is a plant, using ideas from the class "Plants/Animals" chart in their reasons. For example, Jesse's team wrote that they thought that *Daphnia* was an animal, more specifically, an "incect," because "it has little legs and they are moving around." Shawna's group wrote that *Anacharis* was a plant because "a plant has seeds it has bock [bark] and a plant has leaves and som plant have barees [berries]." Two other responses are shown in Figures 3-1 and 3-2.

   At the end of the unit, Mrs. Schwartz reflected on how it had worked in her class. She commented that she had asked the students to carry out both individual and group assessments throughout the unit, and she realized that both were helpful in her teaching. "The individual notebook helps me assess how they are doing on their own. They need to be able to do some things individually." The individual notebooks as well as her interactions with students in class helped Mrs. Schwartz judge their understanding and their readiness for the next activity.

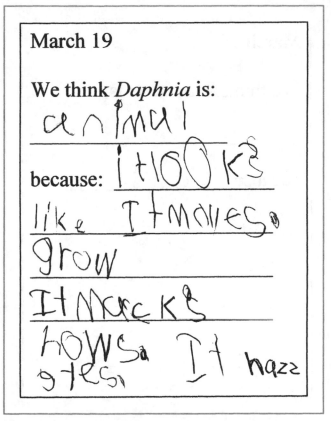

March 19

We think *Daphnia* is:

ca n imal

because: I tlOOK

like ITmoves.

grow

It mrcks

hows It hazz
9tes.

FIG. 3-1 *Justin's Team's Response to the Plant/Animal Identification Game*
Translation: "We think *Daphnia* is animal because it looks like.
It moves. Grows. It makes noise. It has eyes."

"The group notebook helps them to decide together. I had to guide them to talk and decide together. You can't just expect that to happen." Mrs. Schwartz worked hard to develop a classroom community that valued thinking and talking together. The group recording devices were one way to reinforce that students need to work together to solve some problems. However, the students needed constant guidance. Mrs. Schwartz commonly directed students to "Talk to your group," as she circulated among the teams.

The assessment tools that Mrs. Schwartz developed for this unit provided feedback about student progress but also helped achieve some of the social skills goals that she had set for this first-grade class. In the end, she decided that the unit had been worthwhile in helping students work together to make claims

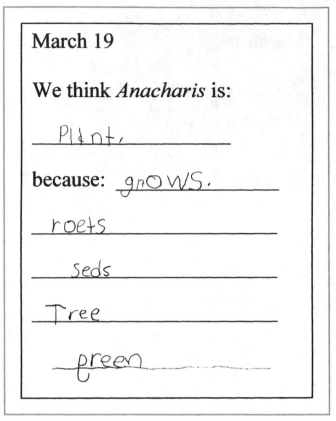

FIG. 3-2 *Elizabeth's Team's Response to the Plant/Animal Identification Game*
Translation: "We think *Anacharis* is plant because grows. Roots. Seeds. Tree. Green."

about the differences between plants and animals using evidence from their observations. Information from her seamless assessment gave her the confidence to feel that way.

Note

1 Planting the seeds in plastic bag gardens was necessary for two reasons. First, students could observe easily when the seeds sprouted. More importantly, they recognized that plants need only water to sprout, which was key for the subsequent investigation in the "Seeds and Eggs" unit.

It All Goes Back to Plants

■ *Michele Lee*

Unit Notes

Grade: 4

Learning Goals: Students will understand that some animals eat plants and some animals eat animals, but almost all animal food can be traced back to plants.

National Science Education Standard: Content Standard: K–4, Life Science: All animals depend on plants. Some animals eat plants for food. Other animals eat animals that eat the plants. (National Research Council 1996, 129)

Assessment Strategies:

  Engage: group discussion with teacher-directed questions

  Explore: collaborative chart

  Explain: exit ticket

  Elaborate: team report

  Evaluate: scenario exam

Vignette

Engage

  "What are those for?"
  "Are we having a party?"
  "Why are there plants on the tables?"

I was greeted with these questions as my fourth graders traipsed into the science classroom and settled into a circle on the floor. Prior to their arrival, I had placed living plants, hay, and sticks in the center of the demonstration table. The students' curiosity led naturally into my first assessment question: "Why do you think animals need plants?"

The discussion was lively as students suggested examples of animals eating plants or using them for their homes. Sarah said, "My guinea pig loves carrots."

Emilio pointed toward the plants on the table as he said, "Cows eat grass and hay."

Billy volunteered, "Birds live in trees."

"We feed Ms. Mitchell's rabbit lettuce sometimes. He loves it!" offered Shalini.

"Pandas eat bamboo," Kelly said.

Piggybacking on this idea, Jerry said, "I saw the giraffes at the zoo eating leaves off trees."

Darnetta claimed, "Squirrels make nests with leaves."

Matthew excitedly pointed out, "Izod eats kale." (Izod was the class iguana.)

From their answers, I gauged students' prior understanding, which predominantly focused on knowledge about pets, farm animals, or animals at the local zoo or in the local environment.

Students did not mention humans in their discussion of plant-animal relations. This did not surprise me because elementary students often do not think of humans as animals (Driver et al. 1994). Thus, I probed further. "Why do you think we—people—need plants?" There was a pause until the student who thought we might be having a party mentioned "decoration." I nodded but looked around expectantly for other responses. I pulled an apple out of my pocket and set it on the demonstration table without saying anything. Picking up on this silent cue, students brainstormed various plants (mostly fruits and vegetables) that people eat.

Now that we had a lot of ideas and examples, I asked the students to vote: "How many of you think that animals need plants? How many of you think people need plants?" The entire class raised their hands for both questions. When I asked why, the explanations included "Animals need to eat them" and "People use plants for food." I intentionally pressed further, asking students to consider animals such as fish, worms, insects, and spiders. A debate arose regarding whether fish flakes were made of plants or not as well as whether an earthworm that eats soil really needed plants. Students were unclear what ants might eat and seemed puzzled when flies were the only food source that they could suggest for spiders. I asked, "Suppose all of the plants in the environment died and disappeared. Would all animals be able to survive?" Some responded with a definitive yes or no, while others were ambivalent. I told the students that we were going to consider the importance of plants. "Our main question to explore further is: Do all animals, including humans, rely on plants?"

Explore

From the discussion, it was clear that students realized that people eat plants. Whether they realized that "almost all kinds of animals' food can be traced back to plants" (American Association for the Advancement of Science 1993, 119) was less obvious. I invented an activity to help them explore this idea. Students went to their team tables, where I had placed a bag of common food items (pictures, toys, or real) derived from plants, animals, or both (e.g., french fries, steak, salad, orange, hot dog, fried chicken, taco, pizza, rice, green beans). Working in groups, students discussed if the food came from a plant or not. Each group recorded responses for two items on a collaborative chart on the whiteboard at the front of the classroom. Once the groups finished, I reconvened the class to examine the collaborative chart and discuss whether people, as animals, eat plants. Everyone agreed. Students were quick to clarify that humans also eat animals. Matthew pointed out that not all humans eat animals. "My neighbor is a vegetarian." I asked him what he meant in case other students did not know that a vegetarian is a person whose food source is predominantly, if not solely, plants. Some students nodded in understanding, while others looked surprised.

After establishing a general consensus that humans use both plants and animals for food, I asked, "What if you decided to eat animals only? Let's say you eat only chickens and cows. Would it be OK if there were no plants?"

Ashley piped up immediately, "No, because plants give us oxygen."

I replied, "That is an excellent point. Plants do produce oxygen that animals need to breathe and live. Let's assume that somehow we are able to get the oxygen we need without plants. Do you think we still would be able to survive without plants as a food source?" I asked students to commit to an answer—yes, no, or maybe/unsure—with a show of thumbs-up, thumbs-down, or thumbs-sideways. The response was mixed. Students who said yes thought that people could survive just by eating meat. The reasons given for no were that "people have to eat plants" or "chickens and cows wouldn't be able to survive without plants." I asked students who gave the latter response to explain.

Wayne said, "Without plants, the animals would have nothing to eat and would die, so people would die because they would have nothing to eat."

Explain

During the *Explain* phase, I wanted to help students understand two important concepts: (1) "some animals eat plants for food" (National Research Council 1996,

129) and (2) some "animals eat animals that eat the plants" (National Research Council 1996, 129). To help students visualize these relations, I made signs that could be worn as necklaces, such as animal (picture of human), animal (picture of a cow and a steak), plant (picture of grass), and plant (picture of corn).

I asked the cow and the corn to come to the front of the class. "How are cows and corn related?" I asked. Students explained that cows eat corn. I drew this relation on the board, linking the animal and the plant with lines. (I drew lines to show the relationship instead of using arrows, which I had read could confuse students [Driver et al. 1994]). We also considered the cow and its relation to other plants such as grass and soy beans, and I added these to the model on the board.

Next I invited the student wearing the "human" label to join the group. Audience members noted that the human ate the cow, which ate the grass and the corn; the performing students held hands to demonstrate how they were connected. I added to our model on the board. "What would happen if the plants died?" I wondered aloud. The corn and the grass dramatically released their hands and fell over "dead"; the human and the cow did the same. I asked another student, representing a chicken, to join the group. She held hands with the human and the corn, but not with the cow, and I added these relationships to our growing model. One by one I invited students to join the group and find plants and/or animals to which they were connected.

It was at this point that I revisited the notion expressed earlier—that people can survive just by eating meat. "What do you think now? Could we really survive if there were no plants to eat?" I asked students to write down one thing that they ate for dinner the night before and link it back to plants. At the conclusion of this lesson, I passed out an exit ticket that asked students to respond to the following prompt: "Almost all kinds of animals' food can be traced back to plants. Do you agree or disagree? Why do you think so? Draw an example to support your answer." All students could draw the relationship between plant eaters and plants, and many included meat eaters in their drawings. Luke's response (see Figure 3-3) demonstrated that he was able to explain the links between both meat-eating and plant-eating animals and plants.

Elaborate

It seemed that students understood the importance of plants in relation to humans and many other animals and could trace relations of some animals back to plants. However, we had not addressed the idea that most other animals' food linked to plants as well. Students initially had been confused about whether the food that animals such as fish, worms, insects, and spiders ate

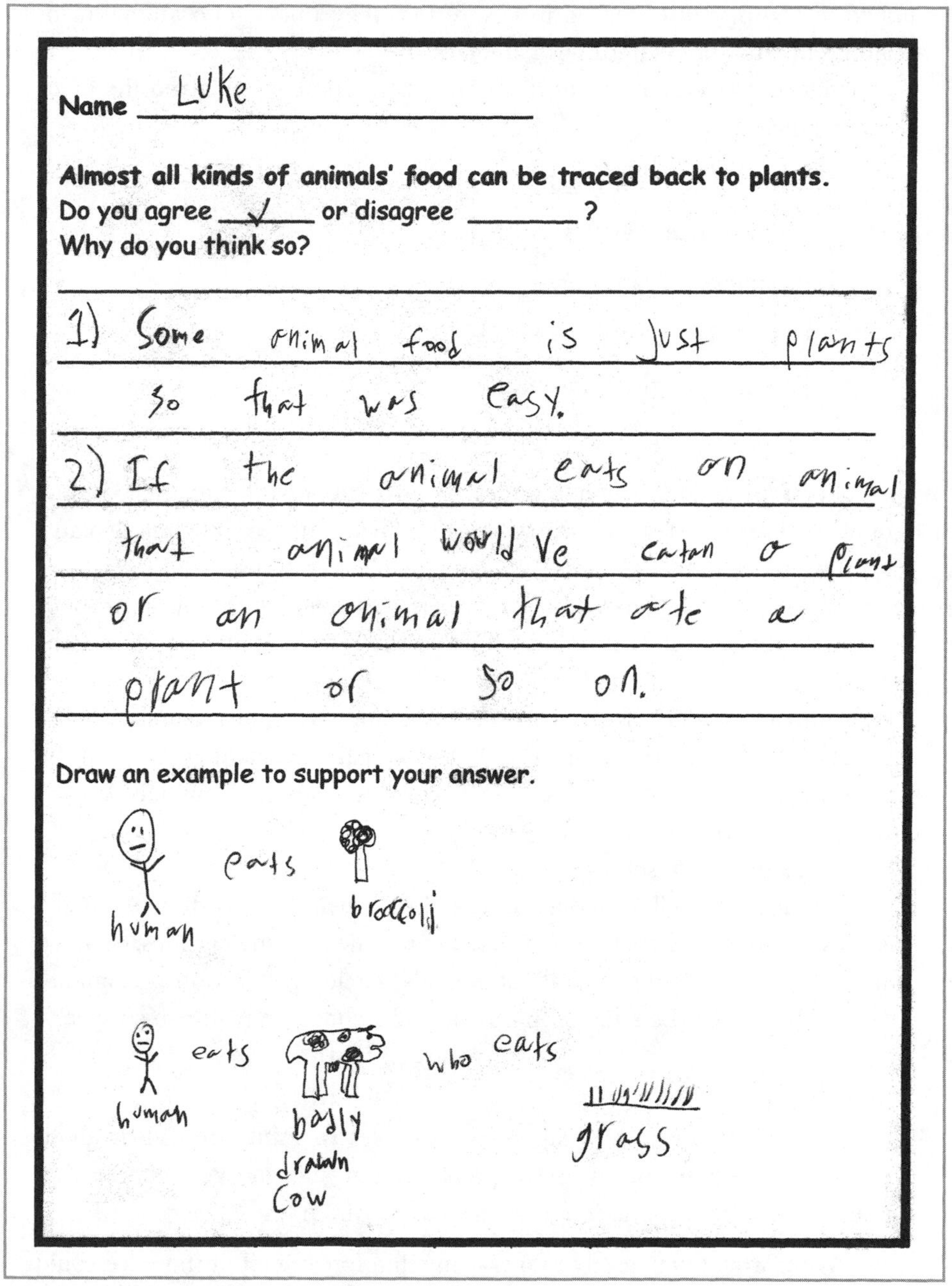

Name __LUKe__

Almost all kinds of animals' food can be traced back to plants.
Do you agree __✓__ or disagree _____?
Why do you think so?

1) Some animal food is just plants so that was easy.

2) If the animal eats an animal that animal would've eaten a plant or an animal that ate a plant or so on.

Draw an example to support your answer.

human eats broccoli

human eats badly drawn cow who eats grass

FIG. 3-3 *Luke's Exit Ticket*

linked to plants, and I wanted to ensure that they understood animal-plant relations broadly, not just for farm animals.

I assigned each student team to research a group of animals and the foods the group might eat:

- pets (fish, dogs, cats, birds)

- sea animals (fish, whales, clams)

- forest animals (lizards, deer, coyotes)

- river/pond animals (fish, mussels, snails)

- soil animals (earthworms, beetles, termites)

- other common insects and spiders

Students used library and Internet resources to investigate their animal group's food sources. Teams discussed how the food sources could link to plants and drew models of the relations. When the whole class reconvened, each team reported their conclusions about how their particular animal group's foods linked to plants, and they supported their conclusions with at least two examples. After each team reported, the class decided where to place each researched animal into a Venn diagram on the whiteboard: animals that eat plants, animals that eat animals, and in the middle, animals that eat plants and animals. At the end of the reporting, we examined the circles. "Are there any animals that did not fall into the circles?" I asked.

Students unanimously responded, "No!"

"What does that tell us about plants and animals?" I asked students to discuss this question with their groups. As I interacted with small groups, I probed students about the "animals that eat animals" circle. Each group was able to articulate how even the animals in this circle depended on plants to survive.

Evaluate

I wanted students to go beyond giving examples to using their ideas about plant-animal relations to solve some problems. I developed two scenarios that I asked students to respond to individually as a written test.

- What would happen to an owl living in a forest if all of the mice could not find food? Why?

- What would happen to the bees in an area if all the flowering plants died? Why?

The test directed students to draw and label a diagram to support their written explanations.

After students completed the test, we discussed their answers. Because in previous lessons students had considered the food relationship between animals and plants, they were able to complete this final evaluation task easily. I asked the class if someone was willing to share what he or she had learned. Eric, one of the quieter students, probably summed it up best: "It all goes back to plants. Even if you don't eat plants, we all still are connected to them because what we eat might eat plants."

What's the Anther?

■ *Marsha Tyson and Kelly Turnbough*

Unit Notes

Grade: 6

Learning Goals: Students will understand that plants and animals (1) reproduce sexually, (2) have comparable reproductive structures and functions, (3) produce egg and sperm cells (gametes), and (4) produce offspring (zygotes) when gametes unite.

National Science Education Standard: Content Standard: 5–8, Life Science: In many species, including humans, females produce eggs and males produce sperm. Plants also reproduce sexually—the egg and the sperm are produced in the flowers of flowering plants. An egg and a sperm unite to begin development of a new individual. That new individual receives genetic information from its mother (via the egg) and its father (via the sperm). Sexually produced offspring never are identical to either of their parents (National Research Council 1996, 157).

Assessment Strategies:

> *Engage*: field trip notebook; discussion
>
> *Explore*: Fast Plant drawings; poster
>
> *Explain*: diagram labeling
>
> *Elaborate*: think, pair, share; Venn diagram intersection
>
> *Evaluate*: cycle diagram; student-generated Venn diagrams

Vignette

Engage

"Do you remember the focus question for today's visit to the nature center?" It was late April and the weather had cooperated with sunny skies and cool temperatures. We had decided the best way to begin the study of reproduction was to organize a field trip to the local nature center to raise student interest in plant reproduction. Upon our arrival at the nature center, we repeated the field trip challenge: "How many different kinds of seeds can you find?"

The student teams were excited and responded enthusiastically, "We'll find the most!" and "We'll get more than you!" and "We'll find enough to make a new nature center!"

To help students remember their roles as budding scientists, we asked, "How do scientists make observations?" This had the expected calming effect.

Pam said, "They look with these," holding up a hand lens.

Steve remarked, "They write stuff down."

Jenni added, "They might make drawings." Armed with hand lenses, notebooks, and pencils, we headed for the woods.

The assignment was for students to find as many types of seeds as they could (at least three), draw each seed, draw the parent plant, and describe how they thought the plant produced its seeds. Adult chaperones monitored each small group of students, who searched for the bounty. Prior to the trip, we had instructed chaperones on how to help students by sharing tips on where and how to look for seeds. We suggested they look in areas with dried brush left over from last fall and look for plants that may be producing flowers and fruits at the same time, such as the Mayapple. We asked chaperones to allow for student discovery with minimal adult guidance.

Back in the classroom, students shared their drawings and their ideas about plant reproduction. We learned that students were aware that flowers played a role in reproduction and that insects (especially bees) were involved in the reproductive process. We also learned that students lacked the language to talk deeply about these concepts. Without appropriate terms, students had difficulty distinguishing structures. For example, students found Mayapples in various stages of reproduction (from early flowers to developing fruit) but were unable to distinguish various parts such as the filament, the anther, and pollen (in association with the stamen) or the style, the stigma, and the ovary (in association with the pistil). In an effort to perpetuate the excitement generated by the field trip, we decided it was time to explore firsthand how seeds are produced. We would explore the concept of sexual reproduction using Wisconsin Fast Plants.

Explore

Sixteen days prior to our study of reproduction, teams of students planted, watered, and recorded the growth and development of a set of Wisconsin Fast Plants.[1] These unique plants complete an entire life cycle in thirty-five to forty-five days (seed → germination → growth → flowering → pollination → fruit → seed production). We had timed the Fast Plants to bloom subsequent to our field trip. We used these plants to help students identify which parts were male and which were female (we identified the stamen as the male part that produced the pollen and the pistil as the female part that produced the ovary and the eggs) and to teach associated terms. Students examined the Wisconsin Fast Plant flowers with their hand lenses and drew the structures in their science notebooks. Aided by black-line diagrams (supplied with the Fast Plants by Carolina Biological Supply Company), students were able to label the male and the female flower parts with little trouble.

To help students think about the roles that these parts played in the production of seeds, we initiated student-led investigations using the Fast Plants by doing a variables scan (Jelly 2001) of the parts they might be able to manipulate in their studies. This activity helped students decide what variable to change and what variable to observe as a result. For example, Nathan, Pam, and Jenni selected the pistil as the variable to change and seed production as the variable to observe. They claimed, "If the pistil is removed, seeds will not be produced." In order to test this claim, they removed the pistils from five plants and left another five intact. They predicted that the plants without pistils would not produce seeds, while the intact plants would. Suzy, Pedro, and Ricky selected pollen as the variable to change and seed production as the variable to observe. They claimed that the pollen from the anther was necessary for seed production. In order to test this claim, they decided to help pollen move from one plant to another by touching cotton swabs to the anthers of five flowers and then touching the stigmas of five others and vice versa. Other groups made similar claims and devised interesting strategies to test them.

Three weeks later, students harvested the seeds. They were excited to find evidence to support or refute their claims. Each group shared their results on a poster listing their claim, evidence, and conclusion. For example, Nathan, Pam, and Jenni wrote:

Claim: No reproduction without pistils.
Evidence: 5 plants without pistils did not produce seeds and 4 out of 5 plants with pistils did produce seeds.
Conclusion: Pistils (females) are needed for seed production.

Suzy, Pedro, and Ricky learned that pollen can be transferred by touching. However, they had not used an experimental control, so it was difficult for them to be confident about the role of touch in pollination. Other students learned that pollen goes from anther to stigma and not from stigma to anther, that wind supports pollination, and that seeds develop below the blossom.

Based on the poster assessment, we realized that students understood that reproduction in these plants required male and female structures (goal 1) and that the male structures produced material that was transferred to the female parts (goal 2). We hoped that some of the students understood that the union of the male parts (cells) with the female parts (cells) was the basis of sexual reproduction (goal 3) and that a seed would be produced from that union (goal 4). However, we knew these concepts would need more direct attention. We decided that we could help students use the evidence from their plant investigations to understand scientific explanations of plant reproduction. We also decided that students were ready to compare plant and animal reproduction to better understand both.

Explain

We drew a Venn diagram on the board and told students we would use the diagram to record characteristics of plant reproduction, characteristics of animal reproduction, and characteristics common to both. We began with plants. "What do we know about plant reproduction from our investigations?" we asked. As students volunteered characteristics of plant reproduction, we summarized and recorded them in the left-hand circle. For example, we wrote, "Some plants have male parts, some have female parts, and some have both male and female flower parts." We also wrote, "Female parts produce eggs and male parts produce pollen." Finally, based on a comment from Ricky, we wrote, "The pollen must join with an egg to produce seeds."

At this point we felt that students were ready to hear a formal summary of plant reproduction and to make the bridge to human reproduction. We commented, "All of these steps in plants are known as sexual reproduction, because there are male and female contributions to making the offspring. Many types of plants reproduce this way, but some do not." We wrote the term *Sexual Reproduction* above the Venn diagram and continued. "Do you think plant and human reproduction is alike in any way?" Some students vehemently shook their heads in opposition to this idea, but a few had quizzical looks on their faces, as if seriously contemplating this rather odd idea.

To help students and ourselves prepare for a discussion of human reproduction, we asked students to write down one claim and one question about human reproduction and turn their papers in without names. This opened the door for many students to share information they had learned from parents or older siblings and to raise questions about human reproduction based on our study of plant reproduction. The next day, we read what students had written aloud to the class, using our judgment to monitor language and content, and taking care to preserve anonymity. We provided black-line diagrams showing human male and female reproductive systems. As we read student questions and provided responses, students labeled their drawings and wrote explanations below each diagram while we added ideas to the right-hand circle on the board. After we had read all the students' questions and claims, we were ready to tackle the intersecting space of the Venn diagram.

Elaborate

We introduced the *Elaborate* phase with a challenge. "Earlier we asked if you thought plant and human reproduction were alike in any way. Let's see what you think now. Write one reproductive structure or function that plants and humans share."

"Are you saying plants and humans have the same parts?" asked Pedro.

"I don't have any seeds," Nathan quipped.

Suzy quickly responded, "A seed is just a baby inside its mother plant, just like babies can grow inside of us." Suzy's comment motivated students to get to work.

After the students had some individual *think* time, we asked them to *pair* up with someone and *share* their ideas.[2] As we circulated, we heard students make a number of statements comparing plant and human reproduction. For example, Nathan said, "Plants and animals have eggs."

Pedro claimed, "Plants have pollen and animals have sperm."

Suzy interjected, "Plants have both male and female parts and humans have one or the other."

After the pairs met, we asked them to share their ideas with the entire class. In addition to the ideas we heard when we circulated, other claims emerged, for example, "Plants and humans both produce offspring" and "Offspring look kind of like their parents." As students presented their ideas, we listed them in the intersecting space of the Venn diagram. Students were surprised to see that the intersecting space representing shared characteristics of plant and animal sexual reproduction contained so many features. This interchange also helped students feel free to ask questions, such as: "Why do plants depend on the wind to aid with fertilization and humans do not?" "Do plants have sex?" "Do plants get pregnant?" and "Do all animals depend on fertilization?"

Evaluate

To reach closure on this unit about reproduction, we directed students to complete an individual summative evaluation. "In your science notebook, draw the stages of plant reproduction in a circular diagram. (Think back to those we drew for the water cycle.) Outside that circle, write notes to describe where human reproduction is similar and different." Most of our students were eager to demonstrate their knowledge and had little difficulty drawing the inner circle. Many could also draw the outer circle for humans. Their products demonstrated that we had achieved the goals of the unit and were ready to move on.

Our final activity of the plant reproduction unit would lead us into the subsequent unit, genetics. We asked students to observe the plants that had grown from the Fast Plant seeds and compare the characteristics of the offspring with characteristics of the parents using a three-circle Venn diagram. In one circle student teams described characteristics in common with the father (pollen provider), in another they described characteristics in common with the mother (ovary provider), and in the third they described characteristics of the offspring. They filled the intersection with characteristics shared by all three (mother, father, offspring).

Some students expressed an interest in making similar diagrams showing relationships within their families (with mothers, fathers, sisters, and brothers). This is a powerful activity that would meet the goals of this instructional sequence. However, instead of asking students to diagram their nuclear family characteristics, we provided a menu of options. Students could choose to diagram their nuclear family, a relationship with a relative, or a relationship with a significant other who was not related (a friend, a guardian, or a teacher). We gave these options because many of our students come from blended or nontraditional families where bloodline relationships are sometimes difficult to trace. We have found that all students enjoy investigating these relationships, in part because doing so accentuates a connection with someone who is close or with whom they hope to build closeness. These relational diagrams would become the basis of our genetics study.

Throughout the reproduction unit, seamless assessment had informed us about student understanding of plant reproduction. It had also helped students ask their own burning questions about human reproduction. Finally, seamless assessment demonstrated to us student readiness to study new concepts that would be addressed in upcoming units.

Notes

1 Wisconsin Fast Plants and accompanying materials are available from Carolina Biological Supply Company.

2 Think, pair, share is a common strategy we use to generate ideas in class. More information on the strategy can be found in Victor and Kellough (2000) and in Chapter 2.

Vignette

Water You Know

■ *Sara Torres*

Unit Notes

Grade: 7

Learning Goals: Students will understand (1) that biotic and abiotic factors affect water quality; (2) that these factors can be related to each other; and (3) that monitoring water quality factors in comparison with water quality standards is important to human health.

National Science Education Standard: Content Standard, 5–8: Science in Personal and Social Perspectives, Personal Health: Natural environments may contain substances (for example, radon and lead) that are harmful to human beings. Maintaining environmental health involves establishing or monitoring quality standards related to use of soil, water, and air. (National Research Council 1996, 168)

Assessment Strategies:

   Engage: water meter

   Explore: water quality survey

   Explain: water quality table and class booklet

   Elaborate: eco-mystery

   Evaluate: conference presentation

Note: This vignette is based on Sara's experience teaching middle school science in the Chihuahuan Desert in southern New Mexico.

Vignette

Engage

"How much water do you think your family uses in one day?" I asked the seventh graders. They brainstormed their families' uses of water, including taking showers, flushing the toilet, washing clothes and dishes and cars, watering the lawn, playing water sports, cooking, and drinking. Using the "Water Meter" activity from Project WET (Project WET Water Education for Teachers 1995), students determined approximately how much water their families typically used in a day. After students made their water-use estimates, I asked another question: "How much rain do you suppose our desert community receives each year?" Students were surprised to find that a desert receives less than eleven inches of rain per year, and they recognized the discrepancy between their water use and precipitation amounts. The "Water Meter" activity engaged students in thinking about an everyday phenomenon and motivated them to learn more.

"If each of us uses this much water a day and we live in a desert where water is limited, where do you think our water comes from?" I asked.

Charlie responded, "From the Rio Grande River."

Veronica chimed in, "From the water plant."

Pablo noted, "Water comes down from the mountains."

Michael added, "The water comes from the arroyos [ditches]."

And Ana claimed, "We get our water from the ground."

I agreed with the students that our water came from all of these places. However, because the most obvious water source in our area was the Rio Grande, I declared that the river would be the center of our study.

Explore

"If our water comes from the Rio Grande, how can we find out if the water is safe to drink, use for cooking, and swim in?" I asked. Students suggested asking a city official or looking the information up on the Internet.

Ana was not satisfied with these ideas. "I think we ought to find out for ourselves." Her suggestion catalyzed our yearlong exploration of the Rio Grande's water quality.

During our next class period, I took students on the first of many field trips we would make to the Rio Grande during the school year. On this first trip, students made observations and took notes about the river. At the end of the visit, they spent time making drawings and writing summaries of what they had observed. They also raised some questions: Why is the river water so brown?

Can we drink the water? What's in those little canals bringing water to the river? Where does the water come from? What lives in the river? and Is the river safe to swim in? These questions would help guide our river study.

Before our next river visit, I divided the student into groups to explore various characteristics of the river. Each group learned to conduct three types of water quality tests: (1) a visual survey, (2) a biotic survey, and (3) a physical, or abiotic, water quality test. For the visual survey, each team became familiar with a protocol that described land use around the river, stream bank conditions, riparian cover, water color and temperature, and weather conditions. For the biotic survey, each team learned to identify and collect samples of macroinvertebrates that might be found in the river. For the abiotic factors, each team became expert on one type of test of physical water quality: discharge (flow), turbidity, nitrates, dissolved oxygen, pH, nitrates, or phosphates.[1]

On our next river trip, teams conducted these water quality tests and recorded their observations and measurements (see Figure 3-4). Back in the classroom, we compiled the visual and biological survey data that each team had collected. To this, we added the physical water quality data and generated

FIG. 3-4 *Students Conduct Water Quality Tests at the Rio Grande*

a profile of the river. Ana, always the thoughtful one, wondered, "But does this mean the water is safe to drink and swim in or not?"

Explain

We would need expert help to interpret our water quality measurements and answer Ana's question. We began on the Internet, searching for information about water quality from the Environmental Protection Agency (EPA) (www .epa.gov) and the New Mexico Environment Department (www.nmenv.state .nm.us/). At these websites students were able to find drinking water quality standards, laws, and agencies responsible for water quality. We used the water quality standards as a baseline against which to compare our data. Each team made a chart that listed the water quality factors that they had researched in the first column, followed by water quality standards in the next column, followed by a column with their data. We found some data that were within the standards and other data that fell outside the standards. In order to summarize our findings, we compiled a water quality booklet made up of our initial river profile and each team's data table.

To understand the significance of our data locally, I invited a water quality engineer from our local water company to join us. She brought along brochures that summarized the annual drinking water quality study for our city, which we compared with our class booklet. As students examined the data, they learned two things. First, they found that our local drinking water met or exceeded all of the EPA's drinking water quality standards. Secondly, they found that some data from the Rio Grande water quality booklet indicated that the river water would not be safe to drink. These findings led to two new questions: How does the water company get the river water clean for us to drink? and Why is the water quality in the river low anyway? The water quality engineer was prepared to answer the first question and invited us to come to the water treatment plant someday for a visit. However, regarding the second question, she only shrugged her shoulders and turned the question back to the students: "Why do you think the water quality is low?" That question led to new activities and further water quality monitoring.

Elaborate

The next class period, I hauled out a big rectangular box filled with soil, sand, and a "river" that would serve as our stream table.[2] "What do you think will happen if it rains on our model?" I asked.

"The river will rise," answered Michael.

"The soil will go into the river," Mia claimed.

As students gathered around the stream table, I sprinkled water onto the land. "Make it rain harder!" Enrique shouted. Because rain in the desert often comes down fast and furious, I made a downpour instead of a drizzle. Sure enough, Mia's and Michael's predictions occurred.

"What do you think would happen if someone fertilized their lawn?" I asked.

Students looked puzzled until Enrique said, "The rain would make some fertilizer go into the river."

"How about if a car leaked oil in the mall parking lot or a factory let some of its wastewater into the stream?" To simulate the release of pollutants, we added food coloring to the ground (blue for the fertilizer, red for the oil, green for the factory) and let it rain some more. Students were amazed to see the food coloring from the various pollution points mingling in the river.

"How do you think our river data will be different after it rains?" I asked. I scheduled our next river visit the day after a big rain. Students noticed that the river was darker, which the turbidity test corroborated. They reasoned that, much like in the stream table, some sediment from the stream bank must have eroded into the river during the rain. Their visual inspection of the stream bank confirmed that hypothesis. Back in the classroom, the dissolved-oxygen group found another big change from their previous data—the level of dissolved oxygen had decreased substantially. The students wondered if these changes were related. We discussed why greater turbidity might affect the level of dissolved oxygen. With my help, students generated two ideas: (1) that the added sediment might "push" the oxygen out the water and (2) that muddier water would absorb more sunlight and be hotter, and hotter water would hold less oxygen. "Do you think we will notice any changes on our biotic survey in the coming weeks?" I asked. Back in their teams, students wrote eco-mysteries[3] about what would happen to the macroinvertebrates, weaving the suspected causes into their stories. Veronica's group wrote a mystery that demonstrated how macroinvertebrate levels decreased when the muddy water blocked the animals' breathing. Charlie's group came to the same result using a different suspect. They thought the increase in water temperature would kill off some of the animals. The eco-mysteries revealed students' understanding about how the biotic and abiotic factors that affected water quality might also be related to each other.

Evaluate

We continued to collect river data throughout the school year. Teams adopted particular areas of the river to study in order to help address the question about why the water quality in some instances was low. They chose sites such as where the irrigation ditches entered the river, above and below the water treatment plant, and above and below a major factory. After our last data-collection period of the school year, each team summarized the data they had collected over the entire period. I supplemented their data with data about their sites collected by students in previous classes. Students also compared these data with data available online from students in other schools further north and south of our location on the Rio Grande. They created spreadsheets, graphs, and charts to demonstrate the changing conditions of their site over time.

By this time, student teams had developed a comprehensive profile of their section of the Rio Grande and had come to understand the importance of environmental monitoring over time. However, they had not had opportunities to communicate what they had learned about the river to other groups. Given the social implications of their work, I thought it was important that their data be shared beyond our classroom. The local university was part of a statewide consortium for environmental education that included higher education, industry, governmental organizations, and the Navajo Nation. One activity of the consortium was to sponsor an Annual Bi-National Water Festival that brought together U.S. and Mexican presenters from nearly three hundred environmental and governmental organizations, as well as students, who gave interactive presentations on such topics as the water cycle, water culture and history, aquifers, and water conservation. This was the perfect venue in which to present our river investigations.

Using the data that they personally had gathered throughout the year, as well as information from other sources, students created PowerPoint presentations about their adopted river spots. Groups practiced giving their presentations to their classmates before attending the water festival. At the conference, teams shared the importance of the tests they conducted and the findings that were generated. They explained biotic and abiotic factors that indicated river water quality. They communicated the outliers in their data and gave possible reasons for them. They shared seasonal differences and tentative explanations. They predicted river water quality in ten years and in one hundred years. Finally, and maybe most importantly, they provided recommendations for ways to improve the quality of the Rio Grande based upon their evidence.

The river study teams were well received at the water festival. It was clear to audience members that the students had worked hard to collect data over time and that they understood the data they had collected and the implications of those data for the community. Through this experience, I learned that I could engage students throughout an entire school year in an investigation of a local issue. Students were motivated to learn because their investigations were meaningful to them and the findings were important to others. Assessment occurred naturally through the investigations that students conducted and the products they created, not artificially through an end-of-unit test. These river watchers had developed the knowledge, skills, and attitudes as science students that would help them become responsible citizens in the future.

Notes

1 Protocols for visual, biotic, and abiotic surveys can be found online at various websites: stream teams (e.g., www.mostreamteam.org/), river watch teams (e.g., http://wildlife.state.co.us/riverwatch/), and the GLOBE Program (www.globe.gov).

2 For ideas on using stream tables, see the *Land and Water* curriculum guide, by the National Science Resources Center (1997).

3 Some years I have asked students to read an eco-mystery, like *Who Really Killed Cock Robin?* (George 1971) or *The Missing 'Gator of Gumbo Limbo* (George 1992), before attempting to write their own.

4 | Seamless Assessment in Physical Science

▪ Introduction

The vignettes in this chapter are based on 5E units of instruction carried out with primary (grades 1–3), intermediate (grades 4–5), and middle-level (grades 6–8) science students. In these units, teachers aimed to develop students' conceptual understanding about force and motion; light, sound, and electricity; and properties of matter. Each vignette illustrates how teachers used seamless assessment to plan and inform their instruction. Table 4-1 describes the grade levels, topics, and assessment strategies used in the vignettes.

Physical science topics form an essential component of the elementary and middle school science program. These topics can be difficult to teach because of the many common misconceptions that students (and sometimes teachers) hold. Physical science topics for elementary and middle-level students, although often abstract, include many opportunities for firsthand experiences with phenomena that fascinate students. Furthermore, studying the physical sciences can help students understand many ideas in the everyday world. That is what the units in this chapter attempt to do.

In the first vignette, Tracy Hager tells about introducing the concept of force to her third graders. She uses everyday experiences with which her students are familiar and helps students build scientific ideas from them. Her story illustrates the kinds of scientific thinking that third graders can do and also contains some ideas that would be transportable to lower grade levels in an introductory forces unit. In addition, because force and motion concepts build throughout the grade levels, this introduction could lead to more complex studies of force and motion in middle school and high school.

**Table 4-1 Physical Science Vignettes by Grade Level,
Topic, and Assessment Strategy**

Vignette	Grade Level	Topic	Assessment Strategies
May the Force Be with You	3	Push, pull, friction	*Engage*: hula hoop and individual Venn diagram card sort *Explore*: station drawings *Explain*: exit sheet *Elaborate*: data table *Evaluate*: constructed-response items
Sounding Off	3	Sound is vibration	*Engage*: group activity and discussion *Explore*: team station worksheets; team tuning fork presentation *Explain*: making a claim; theory choice *Elaborate*: tin can phone drawing *Evaluate*: sound test
Shocking News: Static Electricity	4	Static electricity	*Engage*: demonstration discussion; notebook writing *Explore*: notebook: chart—things that repel and attract *Explain*: notebook: balloons activity *Elaborate*: notebook: static circus descriptions *Evaluate*: static electricity device and description
Mirror, Mirror on the Wall	5	Light	*Engage*: sun and shields diagram *Explore*: demo memo *Explain*: light prediction problem 1 *Elaborate*: light prediction problem 2 *Evaluate*: light prediction problem 3
Will It Float?	8	Density	*Engage*: prediction/observation sheet *Explore*: conclusion writing *Explain*: letter to the teacher *Elaborate*: bowling ball discrepant event *Evaluate*: memo to the boss

In the next vignette, Sandra Abell takes us into another third-grade classroom that has embarked on a study of sound. These students approach the abstract ideas of sound via hands-on activities, minds-on reasoning, and making a model that involves the students themselves as the components. In the unit, the students encounter a big idea in physical science, that the physical world is composed of something that cannot be directly seen but can be transmitted from one substance to another: energy.

In the third vignette, Julie Alexander takes us into a fourth-grade science class where another energy topic is under investigation—static electricity. She helps students experience static electricity in a number of ways before explaining the role of electrons in the process. She takes the abstract ideas underlying electricity and asks students to use them in a fun and challenging application activity. Her unit paves the way for further study of current electricity.

Mark Volkmann, in the next vignette, guides fifth graders in sense making about another form of energy: light. Using everyday materials such as mirrors and flashlights, the fifth graders build their understanding in steps. At each step, their understanding is tested through a motivating challenge. The students see a pattern emerge from their data, and the teacher is able to supply the necessary terminology to solidify their understanding.

Physical science concepts become more difficult when students are asked to relate two different ideas to each other. In the final vignette, Kelly Turnbough helps middle-level students understand that certain properties of matter—mass and volume—have an interesting relationship. The mass of an object compared with its volume is a never-changing property of that object called density. Furthermore, density is a property that a student can use to predict whether an object will sink or float in water. The eighth graders in this vignette struggle to put all of the pieces of this complicated idea together. Their teacher learns how to start with phenomena and ideas and introduce formulas only when students are ready to understand what the formulas represent.

Each physical science unit described in the vignettes was designed and carried out at a particular grade level. However, the *National Science Education Standards*, upon which each unit was built, address grade ranges, not specific grades. We hope you will find ways to adapt these units to your local context and grade level, where age appropriate. We also encourage you to develop new units about other important physical science concepts for elementary and middle-level students, as detailed in the standards within the broad categories of properties of objects, materials, and matter; energy; and force and motion (see National Research Council 1996).

Vignette

May the Force Be with You

■ *Tracy Hager*

Unit Notes

Grade: 3

Learning Goals: Students will understand that (1) a force is a push or pull that can move an object and (2) friction is a force that can stop an object.

National Science Education Standard: Content Standard: K–4, Physical Science: The position and motion of objects can be changed by pushing and pulling. The size of the change is related to the strength of the push or pull. (National Research Council 1996, 127)

Assessment Strategies:

Engage: hula hoop and individual Venn diagram card sort

Explore: station drawing

Explain: exit sheet

Elaborate: data table

Evaluate: constructed-response items

Vignette

I bounced a basketball from the coat closet across the classroom and asked my third graders, "Why do you think the basketball is moving?" Students focused on the role of my hand in their answers.

"Your hand was hitting it!"

"Your hand pushed it to the floor and it bounced back up."

I probed further. "Let's think about the job of a basketball. If I am playing basketball, I want to get the most points by shooting the ball into the basket. Does that require a push or a pull?" After students responded unanimously with "push," I invited them to form a circle on the floor around two overlapping hula hoops, soon to become a Venn diagram. I put the basketball in one hula hoop and placed a small card labeled "push" in the hoop. The "pull" card went in the other hoop. We could now use our Venn diagram to sort a variety of objects according to what type of force it took for the object to do its job.

This launched our third-grade unit on force, a teacher-created curriculum with an accompanying materials kit that circulates in our school district. To introduce the unit, I focused one 5E cycle on the concept that a force is any push or pull that causes things to move. Friction is a force that makes things stop. I used seamless assessment throughout the unit to gauge students' understanding of the concepts and to guide my instruction.

Engage

After identifying the push action that was required for a basketball to do its job, we proceeded to a whole-class sort. I passed out an object and "push" and "pull" cards to each student, who took a few moments to explore his or her object and determine whether a push or a pull was needed for it to do its job.

One at a time students demonstrated how their object worked. After the demonstration, the other children held up the card representing which type of force they thought the object needed to do its job. DeAngelo demonstrated his miniature soccer ball by kicking it lightly. Students showed their "push" cards. Amanda pretended to put on a small sock. Students held up their "pull" cards. Students decided that a timer, a calculator, a screwdriver, and a hole punch use a pushing force. I interrupted their activity to ask how they knew the difference between a push and a pull. Evan replied, "A push means the toy is going away from you and a pull means it is coming toward you."

Jacob had a slingshot. He showed how it worked by launching a small wad of paper. Most students displayed their "pull" cards, but Jacob said that the job of the slingshot was to shoot something, so he thought it would need a pull and then a push. Jacob put the slingshot in the middle, in the intersection of the two hula hoops, to show that it used both forces. Soon other items joined the slingshot in the middle, including a toothbrush, a fork, and a picture of Krista's wheelchair (Krista was a student in our class who had cerebral palsy). (See Figure 4-1.)

FIG. 4-1 *Venn Diagram Activity—Picture of Wheelchair*

"Why did all of these objects move?" I asked.

"We either pushed or pulled them," Gena Marie piped up.

"Objects move because a push or a pull force acts on them," I wrote on our class chart. Then I asked students to use this idea to complete an individual card sort assessment that mirrored our whole-class activity. Students cut out pictures of objects we had not discussed as a class. They glued them onto a Venn diagram paper and wrote an explanation for the type of force required in each picture. Krista explained, "The bowling ball is a push because people push it to knock down the pins."

Maria glued the picture of the wheelbarrow in the center, stating, "When your hands are in back of you, then you are pulling the wheelbarrow. When your hands are in front of you, then you are pushing it." By having students explain their reasoning, I was able to understand their decisions. I have never pulled a wheelbarrow, but Maria's explanation about her experiences justified her placement. Figure 4-2 shows an example of a completed Venn diagram.

Explore

During the *Explore* phase, students rotated among three different force stations, experimented with the materials, and drew in their force journals. At one

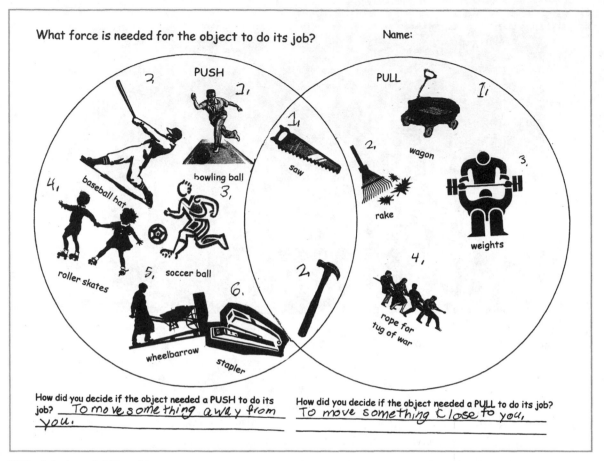

What force is needed for the object to do its job?   Name:

PUSH — 2 baseball bat, 1 bowling ball, 3 soccer ball, 4 roller skates, 5 soccer ball, 6 wheelbarrow, stapler

PULL — 1 wagon, 2 rake, 3 weights, 4 rope for tug of war

saw, hammer

How did you decide if the object needed a PUSH to do its job? __To move something away from you.__

How did you decide if the object needed a PULL to do its job? __To move something close to you,__

FIG. 4-2 *Amanda's Venn Diagram*

station, students used magnets and a variety of objects to find pushes and pulls. At another station students were challenged to move a toy car from a start line to a stop line without touching the car. Students blew the cars, taped a string to the roof and then pulled them from the start to the stop line, and hooked cars together with string and tape. The third station asked students to move a marble. The station included a start line at the top of a ruler propped on books and a stop line at the wall two feet from the ramp. At each station, students drew a diagram to illustrate what they found. They included labels that indicated whether the force was a push or a pull and arrows to demonstrate the direction of the applied force.

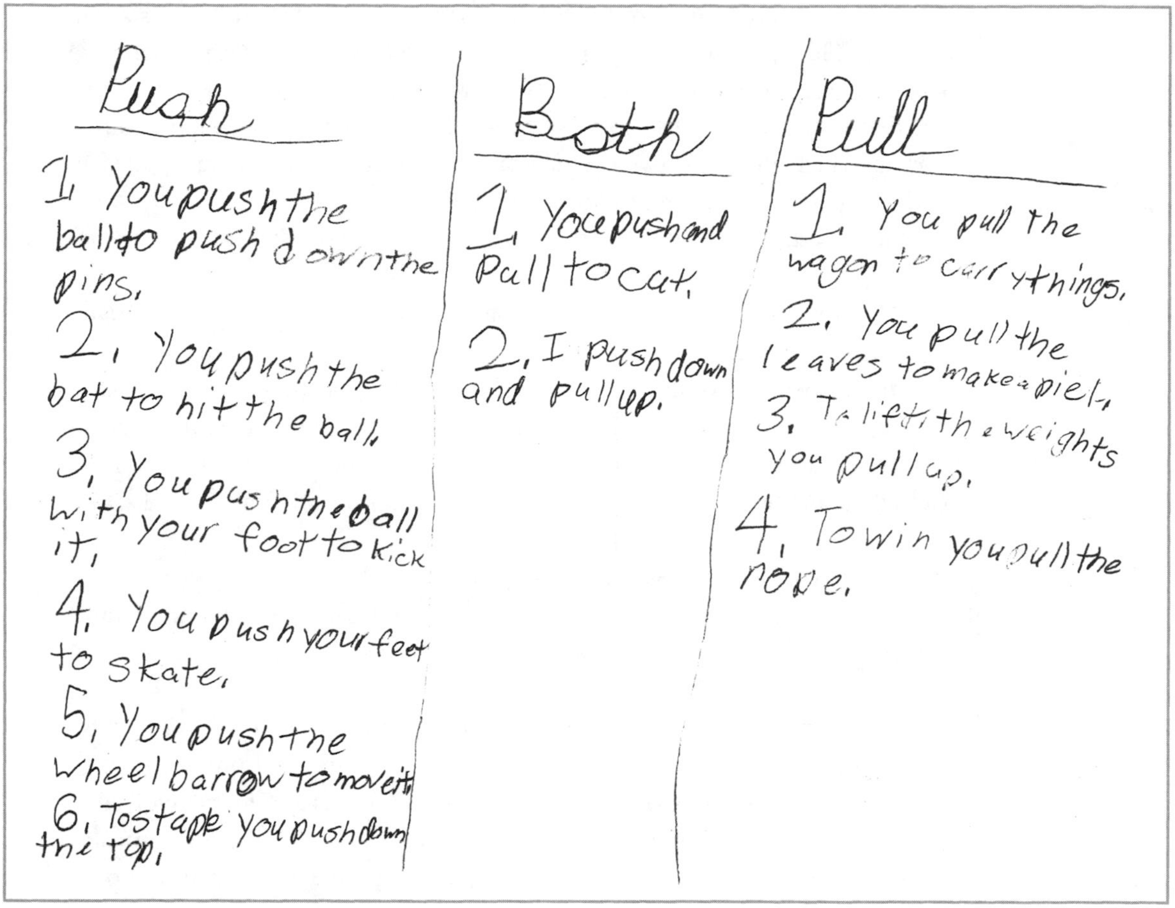

Push

1. You push the ball to push down the pins.

2. You push the bat to hit the ball.

3. You push the ball with your foot to kick it.

4. You push your feet to skate.

5. You push the wheelbarrow to move it.

6. To stop you push down the top.

Both

1. You push and pull to cut.

2. I push down and pull up.

Pull

1. You pull the wagon to carry things.

2. You pull the leaves to make a piel.

3. To lift the weights you pull up.

4. To win you pull the rope.

FIG. 4-2 *Amanda's Venn Diagram (continued)*

Explain

We convened as a class to discuss the station findings. Volunteers demonstrated using the materials from each station. The others used their individual "push" and "pull" cards once more to identify which type of force the volunteer was demonstrating. Savannah pulled two magnets together. She drew her diagram on the chalkboard with an arrow showing the movement of the magnets. She labeled her demonstration as a pull. Many hands shot up as students volunteered to use the same magnets to demonstrate a push. John turned the magnets until one pushed away. Brad commented that the magnets were moving each other. Elizabeth demonstrated using the magnet to pull a paper clip. From their magnet unit in second grade, students remembered exploring pushes and pulls. We labeled those examples with a new card, "magnetic force."

Students were eager to share their findings from the toy car station. Julianna blew her small car. When she drew her example on the board, we labeled it as a push and included the arrows from her mouth to the car to show the direction of the applied force. Next we drew an arrow showing the direction in which the car moved. Emily taped a string to the car and pulled it. "What type of force moved the cars?" I asked.

Julianna said, "A blowing force moved my car from the start line to the stop line."

"My arm pulled my car," Emily commented. I introduced the term *mechanical force* for these examples.

Susanna was our volunteer to demonstrate the marble on the ramp. She gave it a push. "Is there a way to make the marble go down the ramp without pushing it?" I queried. Susanna set it at the top and released it. "What type of force caused the marble to roll down the ramp when Susanna didn't push it?" I asked.

Several students called out, "Gravity."

"What type of a force is gravity—a push or a pull?"

Antoine commented that "gravity was pushing the marble down the ramp." Jacob jumped in and clarified that "gravity is always pulling everything down, toward the center of the earth."

On the chalkboard I listed each type of force: magnetic force, mechanical force, and gravity. I demonstrated additional examples in the front of the classroom. At their desks on exit sheets, students listed the type of force involved in each example and whether it was a push or a pull. From this assessment, students showed they were confident about gravity and magnets but were not as familiar with mechanical force. I would continue to find ways to strengthen their understanding of that concept throughout the unit.

Elaborate

"We know that objects move because a push or a pull force acts on them. Why do you think things stop?" I asked at our next science class. Students offered many descriptions from their previous investigations. "The marble stopped because it hit the wall." "The car stopped because I stopped pulling it." "When the magnet was pulling the other magnet, it stopped because they hit each other. When it was pushing it away, it just couldn't push it anymore."

I retrieved a ramp, books, and a marble and pointed the ramp toward the center of the room with no obstructions. "I wonder what will stop the marble if I roll it toward the center of the room." Austin thought it might hit the far wall. I released the marble and we watched it roll a few feet off of the ramp. "Why do you think it stopped?" I probed.

"It wasn't rolling fast enough to go any farther," suggested Mary.

"But what caused it to stop?" I continued. As students contemplated that question, I asked if the marble would roll farther if we put the ramp in the hall on the tile, or in the grass outside, or on the concrete on the playground.

To find out what causes moving things to stop (a force called *friction*) and to see how different surfaces affect how far something travels before stopping, I provided toy cars, ramps made of plastic gutter material, and a variety of surfaces: sandpaper, waxed paper, plastic grass carpet, fabric, a rubber mat, and plastic wrap. I showed students a setup that included placing one end of the ramp on two dictionaries and the other end on the surface being tested. As a class we listed how to make this experiment a fair test to determine on which surface the car would travel the farthest before stopping. We listed that each team would use the same ramp and ramp height, start the same car at the top of the ramp, and release it (not push it) onto the same variety of surfaces.

Students worked in cooperative teams to predict and test the surfaces placed at the bottom of the ramp. Before conducting their fair tests, the teams ordered the surfaces by how far they predicted the cars would travel on each. They listed this sequence as the first column of their data table. Students placed a car at the top of the ramp and let it roll down the ramp and across each surface. Using a centimeter measuring tape, they measured and recorded the distance the car traveled on each surface (see Figure 4-3). They wrote those distances in a second column of their data table. Some teams performed a second trial using a different car to compare their results. Although the second car traveled a different distance, its relative distance on different surfaces remained the same.

We gathered together to discuss our results. Joe said that he thought the car would "go the farthest on the plastic wrap because it was the thinnest, but it went a lot farther on the waxed paper." Evan's team had had the same prediction. They also found that "the car went farther on the sandpaper than the fabric, which was not what [they had] thought."

"What caused your car to stop moving after it went down the ramp?" I asked. One team referred to the plastic grass surface and said that plastic grass sticking up made it stop. I introduced the term *friction* as a force, and we talked about what that meant. Students recognized there was more friction when the cars moved on rough surfaces like the sandpaper and the plastic grass mat, which caused them to stop sooner. There was less friction when the cars moved on smooth surfaces such as the waxed paper and the plastic wrap, which allowed them to travel farther before stopping. "Friction is a force when two surfaces come together. It is a force that makes things stop," I wrote on our class chart.

FIG. 4-3 *Measuring Distance Traveled by Friction Cars*

Evaluate

Using a "Friction Cars" worksheet and their data tables, students independently answered a series of constructed-response items about their car investigations. (Constructed response is a technique commonly used on our state assessment and one that we practice during our literacy block.) Questions included the following: On which surface did the car travel the greatest distance and why? On which surface did the car travel the shortest distance and why? What forces were acting upon the cars? If we tested the tile floor in the hallway as a surface, would the car stop nearer to the ramp or farther away than it did with the sandpaper? By having students individually complete the constructed-response questions, I was able to judge each child's level of understanding.

Alexis wrote that the car traveled the greatest distance on the "waxed paper because it's smooth, not sticky or rough. . . . The car traveled the shortest distance on the grass because it was very rough and it was plastic." Shelley rationalized that "the waxed paper was thin . . . and the grass was thick." Both girls were beginning to understand that the force of friction is dependent on the texture of the surface. Casey commented that the car went farther on the waxed paper "because it has less friction" and the shortest distance on "the grass because the grass things are sticking up." He was beginning to apply scientific terminology and showed he understood why the car did not travel as far on the plastic grass mat. Other students also used the term *friction* in their answers. For example, Savannah wrote, "the waxed paper had the least friction. . . . The grass had the most friction." Although they did not use the term as precisely as a physicist would, their answers indicated to me that they understood at a level appropriate for a third grader.

The final question required that students predict and apply their ideas to a new, yet similar situation, rolling cars on a tile surface. Every student predicted the car would travel farther on the tile than on the sandpaper. However, the sophistication of the reasons varied from "It would travel a greater distance because it is not all bumpy" to "It has very little friction." (See Figure 4-4.)

Further 5E cycles in the forces unit would explore in greater depth magnetic forces, mechanical forces, and gravity. This introductory cycle established a foundation for further study and helped students build confidence in their own scientific reasoning abilities. Seamless assessment played an important role. It showed me that students understood that pushes and pulls are forces that can cause objects to move, and friction is a force that stops moving objects.

The variety of assessments allowed students different ways to demonstrate their level of understanding. Showing a "push" or a "pull" card during the whole-class sort was a nonthreatening way for every student to give an answer. I could see if they were on track, and they could also see other students' answers and listen to valuable dialogue that scaffolded their thinking. The individual card sort task demanded that they apply what they had learned through the whole-class activity to new objects. During the force stations, students individually recorded the results at each station and then had an opportunity to compare their drawings with classmates' illustrations. To answer constructed-response questions about the friction cars investigation, students needed to draw upon the strategies we had learned during our literacy block and apply push, pull, and friction concepts. They wrote individually after completing investigations in cooperative groups and discussing with the entire class. By gathering information about student understanding in these ways throughout the 5E cycle, I was able to pinpoint misunderstandings immediately and tailor my instruction to meet student needs.

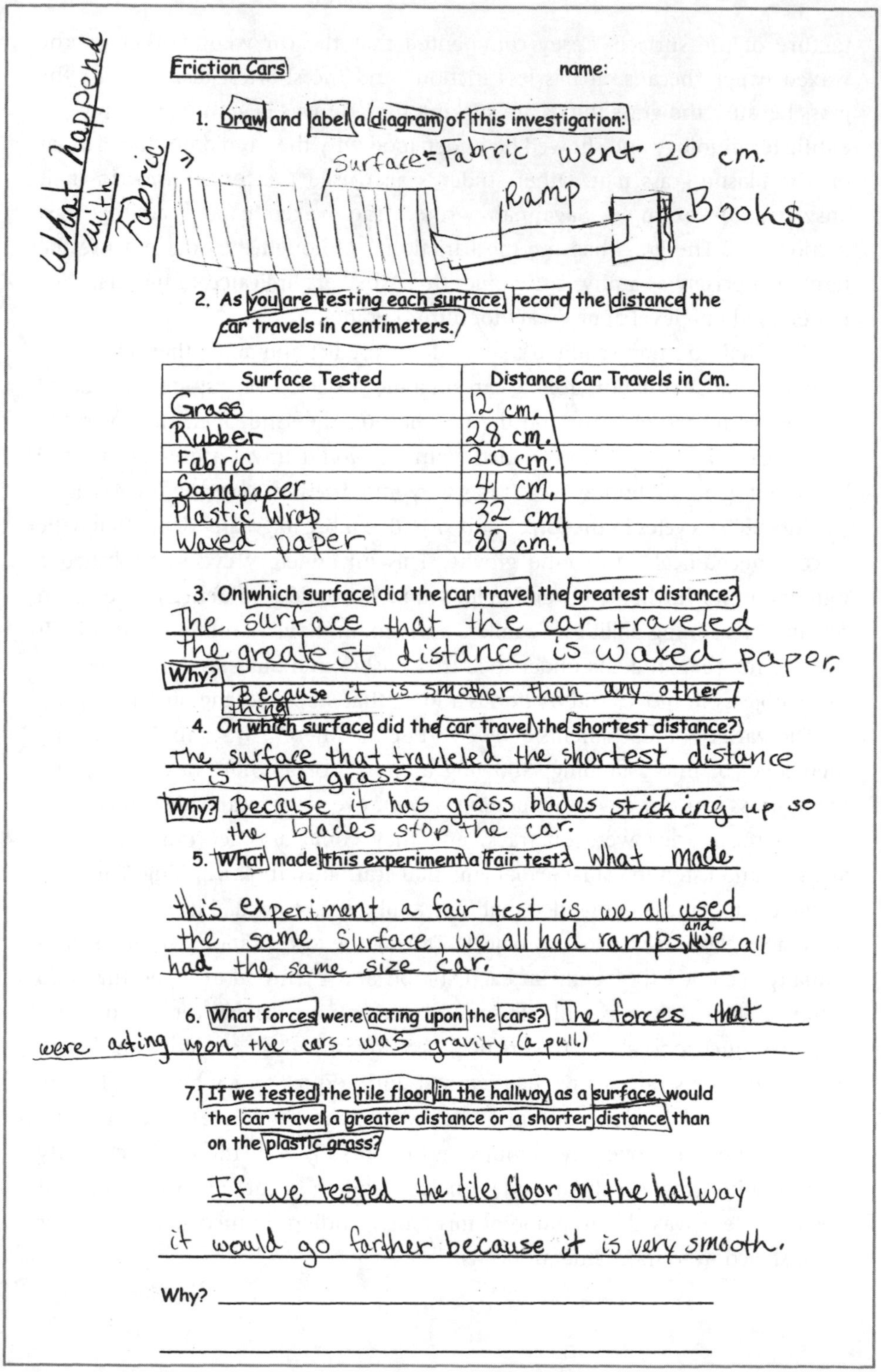

Friction Cars   name:

What happene with Fabric →

1. Draw and label a diagram of this investigation:

Surface=Fabric went 20 cm.
Ramp   Books

2. As you are testing each surface, record the distance the car travels in centimeters.

Surface Tested	Distance Car Travels in Cm.
Grass	12 cm.
Rubber	28 cm.
Fabric	20 cm.
Sandpaper	41 cm.
Plastic Wrap	32 cm.
Waxed paper	80 cm.

3. On which surface did the car travel the greatest distance?
The surface that the car traveled the greatest distance is waxed paper.
Why? Because it is smother than any other thing.

4. On which surface did the car travel the shortest distance?
The surface that traveled the shortest distance is the grass.
Why? Because it has grass blades sticking up so the blades stop the car.

5. What made this experiment a fair test? What made this experiment a fair test is we all used the same surface, we all had ramps, and we all had the same size car.

6. What forces were acting upon the cars? The forces that were acting upon the cars was gravity (a pull)

7. If we tested the tile floor in the hallway as a surface would the car travel a greater distance or a shorter distance than on the plastic grass?

If we tested the tile floor on the hallway it would go farther because it is very smooth.

Why? _____

_____

FIG. 4-4 *Gena Marie's "Friction Cars" Constructed-Response Sheet*

Vignette

Sounding Off

■ *Sandra Abell*

Unit Notes

Grade: 3

Learning Goal: Students will understand that sound is produced by vibration and travels through vibrating solids, liquids, and gases.

National Science Education Standard: Content Standard: K–4, Physical Science: Sound is produced by vibrating objects. (National Research Council 1996, 127)

Assessment Strategies:

   Engage: group activity and discussion

   Explore: team station worksheets; team tuning fork presentation

   Explain: making a claim; theory choice

   Elaborate: tin-can phone drawing

   Evaluate: sound test

Note: Another version of this vignette appeared in Abell, Anderson, and Chezem (2000).

Vignette

The third graders were gathered around their teacher as she sat in a rocking chair, turning pages in Eric Carle's *Very Quiet Cricket* (1990). After reading each page, the teacher asked, "How is the insect making that sound?" Students responded by mimicking some of Carle's words—*rubbing, scraping, munching, slurping,* and *buzzing*—to explain the insects' sound-making activities. Thus began an investigation of sound.

A few years back, I collaborated with two third-grade teachers to design and teach a unit about sound. We designated several learning goals in the unit. For the first part of the unit, we wanted to help students build an understanding that sound is produced by vibrating objects, and later in the unit we would work on the concept of pitch. Our first 5E cycle, and the focus of this vignette, developed the idea of vibration. We used seamless assessment throughout the cycle to help us see how students were building their understanding of this concept.

Engage

After reading *The Very Quiet Cricket,* we invited students to feel their own lips, throat, and nose while making a variety of sounds. When asked to describe what they felt, students used words such as *moving, tickling,* and *vibrating.* These activities served to engage students in the topic of our unit and motivate them for further exploration. They were also assessments. That is, they helped us see that students already had a variety of ways to describe how sounds were made. We decided that students were ready to explore sound in a more focused way.

Explore

For the *Explore* phase, we created a series of stations at which teams of students could interact with a set of materials. The whistles station had straws, metal tubes, and bottles for students to use to make sounds. At the rubber bands station, students plucked rubber bands of varying thicknesses, while at the rulers station, students moved rulers against tables, desks, and chairs to make sounds. The instruments station required students to try out a variety of percussion instruments.

The seamless assessment at each station involved a team worksheet. As teams of four moved through the stations, team members rotated their recording responsibilities. The worksheet questions helped students to focus their observations, for example, How can you make sounds with these items? and What happens to the item when a sound is made? The final worksheet question at each station was aimed at helping students find patterns and make generalizations: How would you describe how you made sounds at this station? After teams rotated through all stations, we reassembled and held a scientists' meeting. Each team demonstrated one sound-making observation and described how the sound was produced. As a result of this session, we learned that many students used words indicating that vibrations make sounds, while others merely described what they had done: "I shook the maracas and they made a sound."

We decided that another explore lesson would help focus students' attention on the connection between sound and vibration. For this exploration, each team received a tuning fork, a small rubber hammer for striking the tuning fork, and a bag of materials (e.g., cup, Ping-Pong ball, plastic dish). We encouraged teams to find different ways to use the materials to make sounds. The assessment device for this explore activity was quite simple. Each team received a chart with two columns: "What We Did" and "What Happened." Within a team, each student tried something with the tuning fork, recorded on the chart, and passed the tuning fork to the next team member. By the end of the exploration, teams had touched the tuning fork to the objects in their bag and to a variety of objects around the classroom.

For the final activity and assessment in the *Explore* phase, we asked teams to demonstrate the most unusual tuning fork experiment they had tried. One team used the overhead projector to demonstrate what happens when a tuning fork touches a dish of water: "It makes it kind of vibrate." Another team noted that they had hit the tuning fork and then put it on the edge of the table, where it "shocked" the table. A third team narrated while demonstrating on one of their members: "They're putting the tuning fork on Sam's glasses and they're making the glasses vibrate."

After the demonstrations, we asked students what patterns they had seen in the data from the stations and the tuning fork demonstrations. Cherril offered, "Almost all of them vibrated." Wanting to be sure that students shared a common meaning for *vibrate*, we probed students for other ways to describe the pattern they had observed. Students used terms such as *movement, moving back and forth*, and *shaking* to describe their observations. At the end of the discussion, we returned to Cherril and asked her to repeat what she had said at the beginning. "Almost all of them vibrate," she stated confidently. Others around the circle nodded in agreement.

This series of learning activities and assessments provided information about our students' understanding of the idea that sound is produced by vibration. Although the students recognized that a variety of objects vibrated, they did not agree that *all* sound is produced by vibration. In the explain phase, we would need to help them refine their thinking and make some cognitive commitments to the concept.

Explain

We started the *Explain* phase by building a model of the human ear. In our model, students played different parts as teachers explained three components of the process: sound making, sound traveling, and sound reception. One student was a drum and another student a drumstick. They produced the vibrations that traveled to the ear. Several students played the role of the ear. Using what they had learned in health class about the anatomy of the ear, one student held a hula hoop covered in plastic wrap that represented the ear drum, and three students became the bones of the middle ear. Beyond the middle ear, one student stretched out on the floor in her role as the cochlea and the auditory nerve. She connected to the last student (who held a picture of a cerebrum), representing the brain. A group of eight more students became air particles, lined up between the drum and the ear. The rest of the students dutifully assumed their role as audience members, knowing that soon the roles would be reversed and they would perform a part of the sound model. Then the simulation began. The drumstick struck the drum, and the drum vibrated. The vibrating drum bumped into an air particle, which bumped into its neighboring air particle and so on down the row until the eardrum was bumped. The vibrating eardrum bumped the first ear bone. One bumped bone led to another and another, until the vibration was converted to a nerve impulse that activated the auditory nerve, sending a signal to the brain, who shouted with glee, "I hear a drum!"

We assessed student understanding at this point by asking each team to make claims about their understanding of sound: "Write three sentences about sound that you can agree about." The resulting sentences mentioned vibrations, sound traveling, and hearing sounds. Their sentences referred to both in-class activities and everyday experiences. Here are some of the sentences:

- Sound can travel through almost anything.

- If you put your ear against the ground you can hear vibration.

- If you put your hand on your throat and talk you can feel vibration.

- It vibrates when it goes through something.

- If you put something close to your ear you could hear it very good.

- It's like a vibration.

- When you touch your Adam's apple and talk you can feel it vibrating.

- When something hits something it makes a sound or it vibrates and makes a sound.

- When you talk the sound vibrates and goes to your ears.

These sentences told us that, to some extent, students understood the role of vibration in producing and hearing sounds. However, thinking back to Cherril's comment that "almost all" sounds were made by vibrations, we decided to probe a bit further. We assessed students' thinking at this point using a "theory choice" strategy. We wrote two sound theories on the board:

- All sounds are caused by vibration.

- Some sounds are caused by vibration and others are not.

We asked students to commit to one of the theories and offer evidence from class activities. After listening to their reasoning, we explained that scientists would accept the first theory as the most accurate. Because not all students were convinced, we realized that they needed further experience with the concept.

Elaborate

A common childhood activity for us as teachers had been making tin-can phones. We were surprised, therefore, when most of our students found this activity new and interesting. In pairs they made and tried out their phones. Then, to assess their understanding, we asked them to draw what they thought was happening to the sound. Student drawings revealed their thinking—almost everyone drew some kind of squiggly line around or emanating from the string to represent sound (see Figure 4-5). Most students called the lines "sound waves." We asked, "Are the sound waves particles of sound going through the string, or are the sound waves vibrations of the string itself?" Six of the nine teams agreed with the more scientifically acceptable idea that sound waves are vibrations (energy), not particles (matter), and could support their choice with evidence from other class activities. We confirmed that scientists would accept that sound was energy in the form of vibrations. At this point, we felt it was time to reach closure on this part of our sound unit and perform a summative assessment.

Evaluate

Up to this point, most of the assessment information we had was based on group discussions and team products. Such information had helped us plan the next steps in our instruction but did not provide information about individual

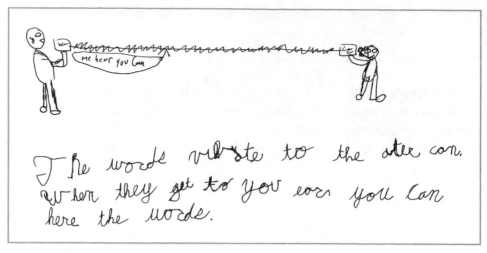

FIG. 4-5 *Third Grader's Sound Drawing*

students' thinking. In order to find out what individuals understood at the end of the unit, we developed a sound test made up of a number of constructed-response items. The first items on the test were simple ones that we knew all students could answer with confidence. For example, What are two things you learned about sound? and What does the word *vibration* mean to you? Subsequent items required higher-level thinking and deeper understanding of the principle of sound and vibration. We tried to put all items into contexts that would be familiar to students. Some of those contexts were similar to class activities. For example, we asked students to design a way to communicate from their tree house to the ground. We asked students to both write and draw, hoping to capture most of the preferred learning modes in our class. We also engaged students in justifying and communicating their reasoning: Your friend says a ball makes a sound when it bounces, but that sound is *not* caused by vibration. Do you agree or disagree? Why? Finally, we asked them to apply their understanding to an everyday situation: Ricky was not hearing very well at school and his ear was hurting. He went to the nurse, who said he had a buildup of wax in his ear between the pinna and the eardrum. Use what you know about how sound travels to explain why the wax buildup would affect his hearing.

We were somewhat disappointed in the results of our testing. Students who were high achievers tended to outperform the traditionally low achievers, even though we had evidence that many of the latter students had understood much about sound. Our seamless assessment throughout the sound unit had informed us in ways that the summative evaluation did not. Our continuing challenge would be to help all students understand science and be able to demonstrate their understanding in multiple ways.

Shocking News: Static Electricity

■ *Julie Alexander*

Unit Notes

Grade: 4

Learning Goals: Students will understand that (1) static electricity is when a body has a net charge, either positive or negative and (2) bodies with like static charges repel and bodies with different static charges attract.

National Science Education Standard: Content Standard: K–4, Physical Science: Students should develop an understanding of light, heat, electricity, and magnetism. (National Research Council 1996, 123)

Assessment Strategies:

Engage: demonstration discussion; notebook writing

Explore: notebook: chart—things that repel and attract

Explain: notebook: balloons activity

Elaborate: notebook: static circus descriptions

Evaluate: static electricity device and description

Vignette

The room was dark, and the children were staring at a large silver ball that sent out jagged sparks. The sounds of crackling and popping were mingled with comments such as "Wow!" "Cool!" "That looks like lightning," and "I bet that would hurt!"

All of this shocking behavior was part of our fourth-grade unit on static electricity. Two of our district's science objectives state that students will demonstrate that they can produce static electricity and that they will be able to describe how electrical charges affect objects. For this introductory part of the unit, I wanted my students to demonstrate their understanding of static charges. Because students are unable to directly observe the movement of electrical charges, and because understanding improves with experience, I wanted to provide a variety of experiences with different objects. What better objects with which to start than the students themselves!

Engage

When the lights came on, students had many questions. "What is that crackling noise?"

"Will that thing hurt if you touch it?"

I asked Melanie to come up and stand on a plastic stool. She put her hand on the metal ball, and I turned on the generator (a Van de Graaff generator[1]). Her hair stood straight up amid roars of laughter. Students commented, "It looks like snakes," and asked, "Why does her hair do that?" I asked Melanie if it hurt, and she said no. I told her to jump down on the count of three and rub her feet on the rug. After a few more students touched the ball and made the class laugh at their spiky hairdos, we began to talk. The students noticed that before each person touched the ball part of the generator, I touched the ball with a small silver sphere on the end of a plastic pole. "Why do you use the wand to touch the ball?" they asked. I told them I used the wand to avoid getting shocked.

"So the electricity goes into the little ball so you don't get shocked," said Zach.

"Exactly! Let's see what else static electricity can do," I replied.

Before we looked further at the static ball, I wanted to know what ideas students had about static electricity. I asked them to write in their notebooks everything they knew about electricity. After a few minutes, they shared what they wrote. Most of the comments the students made were about batteries and current electricity. A few of them wrote they knew about Ben Franklin discovering electricity and that lightning was electricity. After our discussion, I did a few more demonstrations to get them thinking and asking questions to help direct the focus of the unit.

First, I taped long strands of paper towel to the head of the ball. I asked the students to predict what would happen when I turned on the generator. The students correctly predicted that the paper towel "hair" would stand up on the ball. I then asked them how we could take care of this bad hair day, and students said, "Turn it off!" I did, but the paper towel strips continued to stand straight up. The students told me to touch the ball with the wand. When the paper hair fell, I asked the students why it worked. Drew said, "The electricity is still in the hair until the little ball touches the big ball. Then the electricity all goes into the little ball and the hair falls down."

For the second demonstration, I used a Styrofoam block, fur, an aluminum pie plate with a plastic cup glued to its bottom to serve as a handle, and a piece of Christmas tree tinsel tied into a loop. I then demonstrated the floating tinsel halo. After charging the Styrofoam by rubbing it with fur, I charged the pie plate by touching it facedown to the foam block. Next I touched one of my fingers to the pie plate. I then picked up the pie plate by the cup handle and dropped the tinsel loop into the plate. When the tinsel hit the plate, it was repelled, forming a halo floating above the plate. I asked the students what they saw as I moved the pan around. Megan said, "It looked like a force was pushing the tinsel halo up in the air."

"How can I stop the force?" I asked. Kayla suggested that I touch the pie pan. I did, and the halo fell. Students commented that the tinsel and the paper towel hair on the ball acted the same way, because the hair fell when the little ball touched the big one, and the tinsel fell when I touched the pan. I recorded student questions and observations on a class list that we could return to throughout the unit. Their questions included Why does the tinsel float? What is that crackling sound static makes? and Why does your hair stand up?

Explore

After the day of demonstrations, I wanted the students to explore the phenomena of attraction and repulsion. I gave each group several items: a large piece of plastic wrap, a paper towel, a lunch tray, and small plastic bags containing thread, puffed rice cereal, coffee grounds, kitty litter, and pencil shavings. They began the exploration by rubbing the paper towel on the plastic wrap and picking the plastic wrap up in the middle. The edges of the plastic wrap pushed away from each other, making a tent. I asked one student to put her arm under the plastic wrap tent to see what would happen. The edges of the plastic wrap pulled toward the student's arm and clung to it. When she pulled her arm back out of the plastic wrap tent, the plastic wrap pushed away from itself, retuning to the tent shape. The kids loved this and spent several minutes sticking their arms inside the plastic wrap tent and pulling them out, causing the flaps of plastic wrap to move back and forth. After about five minutes, we stopped and talked.

"What do you notice about the plastic wrap?" I asked.

"It pushes away from itself."

"All the time?"

"Only when I pull my arm out."

"Does it do anything else?"

"It sticks to me."

"Does it do this every time?"

"No, the static runs out and you have to put electricity back on it."

"Let's see how other objects respond to the plastic wrap. Use the things you have in your bag and see how they act. In your notebook make a chart to keep track of the things that stick together or push away from the plastic wrap."

The teams poured their items onto the lunch tray. Holding the plastic wrap above the tray, they observed that many items would jump up and stick to the plastic wrap while other items would jump away from the charged wrap. Others observed that when their hands got close to the plastic, many objects would float away from their finger, but still stick to the plastic wrap.

"Why do you think it moves away like that?" I asked. Sarah said it was because the static was all gone. "If the static is all gone," I wondered, "why is the stuff still sticking to the plastic? Shouldn't it fall back to the pan?" The children were not sure about this. "Does this behavior of sticking together and pushing away remind you of anything else you have seen before?" Hands shot up around the room.

Ian responded, "Magnets act like that because the poles stick together and push apart."

"What were the words you used to describe the pushing-apart force of a magnet?" I asked.

"Repel is when they push apart and attract is when they stick together," said Paul. Students agreed that we could use the same words, *repel* and *attract*, to describe the behavior of the items and the plastic wrap. They wrote these words in their notebooks.

Explain

We began the *Explain* lesson by discussing electrical charges and atoms. On the board I drew a planetary model of the atom. In the drawing, the negatively charged electrons circled the nucleus of the atom, and the positively charged protons were in the center of the atom. I told students that positive charges stay with the atom and negative charges can move to other atoms. When negative charges move away from an atom, there is an imbalance of charges, resulting in a positive electric charge to the original atom and a negative electric charge to the new atom. It is this imbalance of negatively and positively charged particles that creates a static electrical charge.

At each table students played a game called "Balance Your Charges." Each student had a balloon-shaped game piece that they moved around the balloon-shaped game board. Each student started out with the same number of positive and negative charges on his or her game piece—therefore, it had no charge. As the players circled the board, they lost or gained negative charges by following the commands on the game board: rubbing their socks on the carpet, getting wet, or rubbing fuzzy fabric. During the game, I walked around and asked students what charges were on their balloons and how a balloon would behave near another balloon at their table.

At the end of the game, students received actual balloons and rubbed them on their heads or clothing. I asked why the balloons were sticking to their heads and shirts. They were able to respond that the balloon had extra negative charges and their hair had extra positive charges, so they stuck together or were attracted to each other. We then charged balloons again and held them close to each other. The balloons pushed away from each other, and students responded that these balloons had the same charge and so pushed away from each other. Students recorded their observations of how charged objects behave, either sticking together or pushing away from each other, in their notebooks. I read these responses and wrote comments and questions on sticky notes, such as "I like the way you explained what you observed," "Do you mean *attraction* instead of *repulsion*?" and "Where did the static charge come from?" My comments were geared toward directing the students to the idea that static charge is a buildup of charge on an object and that objects with like charges will repel, while objects with unlike charges will attract.

Elaborate

After reading their notebooks, I felt students had a fairly good grasp of the concepts of attraction and repulsion, but some needed one more hands-on experience. The next day in class we held a static circus. At three stations around the room, students used common materials to make circus acts, like dancing bears (paper bears on a Styrofoam plate and a tin pie pan), snake charmers (balloons and yarn), and bending water (balloons and water). Paper bears jumped and flipped through the air as students rubbed the pan with their hair or shirt and put paper bears in it. Yarn snakes climbed up in the air toward charged balloons. Streams of water from the sink moved away from charged balloons.

After students interacted with all the materials, each picked one event to write about in his or her science notebook. I asked students to describe what had happened and explain it using what they knew about charges and static electricity. Most students stated in their writing that the static charges caused the items to either attract or repel. I read the notebooks and placed sticky notes with comments on their pages to prepare students for the upcoming performance event.

Evaluate

Most of the assessment information I had collected up to this point was based on class discussion and short individual notebook responses. I needed to get more in-depth information about each student's understanding. Instead of using the district's assessment, which consisted of five questions answered by writing the word *attract* or *repel* (all comprehension-level questions, the lowest on Bloom's taxonomy), I designed a performance assessment that required students to apply their knowledge and synthesize what they knew (higher-order thinking, according to Bloom; see Anderson and Krathwohl 2001). I challenged students to create something that would produce static and explain their results. I did not care if the students used the words *attract* and *repel*, but I was looking for their understanding of what happens when charged objects are brought together. Do they stick together or push away and why?

One student's response is displayed in Figure 4-6. My challenge allowed Sydney to both demonstrate understanding and use her creativity to solve a problem. Although Sydney's example uses the vocabulary *repel* and *attract*, her entry illustrates a much clearer insight into her understanding of the concepts involved than the correct vocabulary alone. Sydney demonstrated her knowledge of static charge by explaining where the charges originated, if they were the same charge or opposite, and how she knew. For this formal assessment, I used a rubric to score students and provide a grade.

Throughout the static electricity unit, I continuously assessed students, formally and informally. Assessment does not have to include pencils and paper all the time, nor does it need to be formally graded. Assessment can occur in the form of questioning and noting responses, listening to students' conversations, or reading what they have written in their notebooks. If teachers want their students to be successful, they will assess students seamlessly throughout a unit, using both formal and informal means to provide meaningful instruction and increase student learning.

Note

1 More information on Van de Graaff generators can be found on the Internet. One fact-filled website is from the Museum of Science in Boston: www.mos.org/sln/toe/history.html.

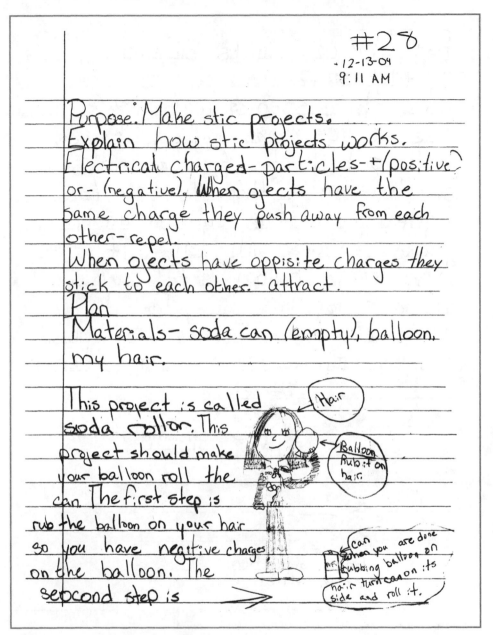

#28
· 12-13-04
9:11 AM

Purpose: Make stic projects.
Explain how stic projects works.
Electrical charged-particles-+(positive)
or- (negative). When ojects have the
same charge they push away from each
other-repel.
When ojects have oppisite charges they
stick to each other.-attract.
Plan
Materials- soda can (empty), balloon,
my hair.

This project is called
soda rollor. This
project should make
your balloon roll the
can. The first step is
rub the balloon on your hair
so you have negitive charges
on the balloon. The
secocond step is →

Hair

Balloon
Rub it on
hair.

can when you are done
rubbing balloon on
hair turn can on its
side and roll it.

FIG. 4-6 *Sydney's Static Electricity Device Description*

Put the can on its side on a flat surface.
The third step is make sure the balloon has enough charges, the put it close to the balloon but don't make them touch. Then lightly move the balloon around and the can will follow it. Since the balloon is following it and not repeling away from it. that means the charges are oppisite and are attracted to each other.

Attract

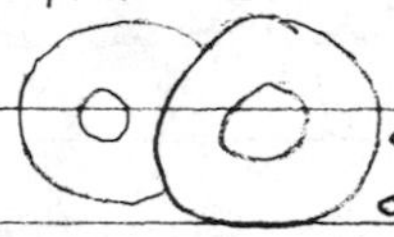

Sticking to each other.

Repel

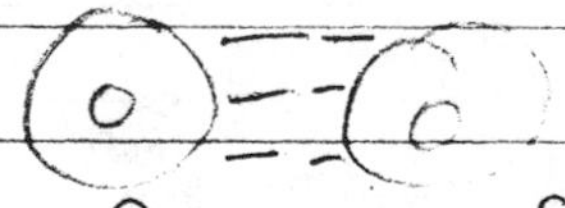

Going away from each other.

FIG. 4-6 *Sydney's Static-Electricity Device Description (continued)*

Vignette

Mirror, Mirror on the Wall

■ *Mark Volkmann*

Unit Notes

Grade: 5

Learning Goal: Students will understand that light reflects off mirrors in a predictable manner governed by the law of reflection. (Law of reflection: Light that strikes a flat and shiny surface will reflect at an angle equal to the angle of the incoming [incident] light ray.)

National Science Education Standard: Content Standard: 5–8, Physical Science: Light interacts with matter by transmission (including refraction), absorption, or scattering (including reflection). To see an object, light from that object—emitted by or scattered from it—must enter the eye.

Assessment Strategies:

   Engage: sun and shields diagram

   Explore: demo memo

   Explain: Light Prediction Problem 1

   Elaborate: Light Prediction Problem 2

   Evaluate: Light Prediction Problem 3

Vignette

Engage

"Does anyone know how Archimedes helped King Hiero II of Sicily defeat the Roman navy in 215 BCE?" (Hakim 2004) I asked to begin our investigation into light.

The fifth graders looked puzzled, and someone even murmured, "I thought this was science, not history."

Unfazed by their comments, I proceeded. "Archimedes arranged shields to reflect sunlight onto the cotton sails of the approaching navy. The light, concentrated by many shields, ignited the sails and defeated the navy before the ships could reach land." The grumbling transformed into comments like "Cool!" and "Wow!" As I told the story, I drew a diagram on the board showing the water, the land, and the ships. I asked the students to copy the diagram into their science notebooks. "Draw where you think the sun and the shields were located to create this solar-powered weapon." I could tell by the rising noise level that the students liked the story and the challenge.

As they completed the diagrams, I noted where they located the sun and how they positioned the shields. Some drew sunrays entering the picture from no distinct direction. Others drew the sun overhead or behind the army. I had reasoned that the sun should be placed behind the approaching navy and the shields positioned and coordinated to reflect sunlight onto each ship, employing the law of reflection. Student diagrams provided me with a wealth of information about the clarity of their thinking related to the behavior of light. Based on this information, I realized that students did not understand how light reflected off shiny, flat surfaces.

Explore

I passed out small mirrors to represent shields and turned the overhead projector to face the students' desks to represent the sun. I challenged the students to reproduce Archimedes' weapon. Working in pairs, students reflected the overhead light from their mirrors onto a toy sailboat placed on a desk in the center of the classroom. I kept a sharp eye out for how well students were able to meet the challenge. Alicia and Elegan were the quickest. With seemingly little effort, they were able to "bounce" light from one mirror to another to the boat. Dylan and Alex had trouble deciding how many mirrors to use and who would hold the mirrors. After several trials, they found a light pathway that worked. After each pair had a turn demonstrating their method to "burn" the sails, I asked each student to write a demo memo that addressed the following questions:

1. How did you position the mirrors (shields) to reflect the light onto the ship?

2. Using your experience with aiming the light, do you think this story about Archimedes could be true? Redraw your initial diagram and explain how this could work.

Students believed that Archimedes had lived and that he had helped Hiero II fight the invaders from Rome. They had fun imagining how the soldiers could have used their shiny shields as futuristic weapons. They were less sure about how to hold the mirrors to reflect the light onto the ships. Their demo memos described soldiers with shields reflecting light. They drew how the light hit the shields and bounced to the target. However, no one represented the critical position of the shields or the angles made by the incoming and reflected rays of light. The demo memo assessment showed me that I needed to help them make their discoveries explicit; that is, I needed to help them observe and explain that the angle of the reflected ray is equal to the angle of the incoming, or incident, ray.

Explain

In our next class period, I asked the students, "What happens when light strikes a mirror?"

Students responded, "It bounces off," and "It travels in a new direction."

I pushed them a little further. "Could you accurately predict every time the direction the light will bounce off the mirror?" I gave each pair of students a flashlight (covered by a mask with a slit), a small metal mirror, a protractor, and a sheet of typing paper. I demonstrated how to use the protractor to measure and record angles and I showed how to trace light rays onto the piece of paper. I did these things in order to direct students' attention to the surface of the mirror and the rays of light coming into and reflecting off the mirror.

After they had some time to explore with these materials, I upped the ante just a bit. I handed out Light Prediction Problem 1 (see Figure 4-7) and took away the flashlights and the mirrors. I challenged the students to be exact in predicting where to place one mirror so light from a flashlight would reflect off the mirror and hit a ball. Each team of two had to draw a prediction before I would return their flashlights and mirrors. Then they tested their ideas. Cries of victory and despair emanated from around the room.

I invited each team to demonstrate their light ray prediction diagram and explain why it did or did not work. As they took turns sharing, I probed them to compare the angle of the incoming ray with the angle of the bounced ray. After a few teams shared, I introduced the terms *incident* for incoming light and

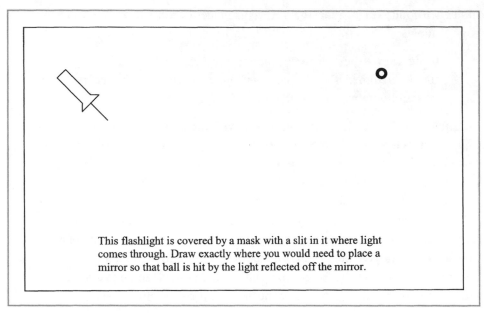

This flashlight is covered by a mask with a slit in it where light comes through. Draw exactly where you would need to place a mirror so that ball is hit by the light reflected off the mirror.

FIG. 4-7 *Light Prediction Problem 1*

reflected for bounced light. These terms made sense to the students at this point because the terms were connected to their experiences. As practice in drawing angles of incidence and angles of reflection, I introduced the *normal line*. A normal line is an imaginary line used in ray diagrams that is drawn perpendicular to the surface of an object at the point where a light ray enters. The use of this term would help students draw accurate ray diagrams now and in future light lessons. Finally, I asked students to develop a rule to predict the path of the reflected ray. By this point, students were quick to see the pattern. "The incident ray and the reflected ray make equal angles," Alice said. A wave of understanding rippled across the class.

"That's what your science book calls the law of reflection," I confirmed. "In order to celebrate Alice's important discovery, we are going to post that law on our 'Laws of Science' chart!"

As a check for understanding, I returned students' initial drawings of shields and ships. "Pick one soldier with a shield and one ship and use the law of reflection to show how that soldier was able to reflect sunlight to hit the sail," I instructed. I wanted them to demonstrate their understanding of the law of reflection by making their drawings scientifically accurate. "I will be using a protractor to measure your changes," I warned. This warning motivated students to put their protractor skills to use once more. With a little coaxing, most students were able to refine their drawings to demonstrate the law of reflection accurately.

Elaborate

To help students deepen their understanding of the law of reflection, I challenged them to solve Light Prediction Problem 2 (see Figure 4-8). I placed a twelve-by-twelve-inch mirror on my desk, facing the class and covered with a sheet of paper. Next, I placed four observers (students A, B, C, and D) in a row across the front of the class. Finally, I placed a flashlight (this time with no mask) as shown in the figure. "Who do you think will be able to see the light from the flashlight? Will it be observer A or B or C or D, or some combination of A, B, C, and D?"

In order to solve this problem, students needed to consider how the light from the flashlight would reflect off the mirror. I asked students, including the observers, to make a prediction. Then I removed the paper covering the mirror and turned on the flashlight. Most students were surprised to learn that the only observer who could see the light in the mirror was student A.

To help them elaborate on their model of the law of reflection, I gave them each a copy of Light Prediction Sheet 2 (see Figure 4-8) and asked them to draw a ray diagram. Most drew a single ray leaving the flashlight, traveling to the mirror, and reflecting at an equal angle to observer A. A few students drew multiple rays from the flashlight to the mirror, each reflected at an angle equal

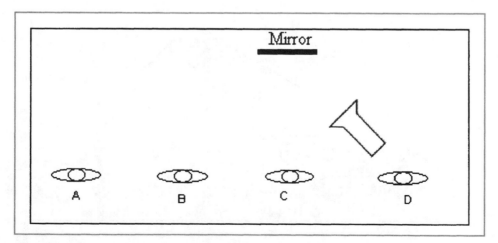

FIG. 4-8 *Light Prediction Problem 2*

to that made by the incident ray. I asked students to take turns sharing their diagrams. We discussed how the multiple-ray diagrams showed an area where any observer could stand and see the light. We took turns investigating this idea by standing within the predicted sight area (see Figure 4-9).

Evaluate

To find out if the students could apply these concepts about reflection, I gave them a real-life problem to solve. "I like to watch the ten o'clock news, but I have a recurring problem. Whenever I sit on my sofa with my living room lamp turned on, I have trouble seeing the TV because of the glare of the lamp. Where should I sit to avoid the reflection of the floor lamp on the television screen?" As I told the students about my problem, I projected a slide showing the layout of my living room on the wall. Several students expressed similar experiences. I gave students a copy of the living room diagram and a protractor. I challenged them to make a claim about the best place to sit and to support their claim with evidence.

Students drew ray diagrams of their solutions and took turns presenting them to the class. Several students used the idea we had developed in the elaborate lesson (see Figure 4-9) to diagram the entire area affected by the lamp. The successful students used their ray diagrams to demonstrate that the upper end of the sofa and the chair were the best places for me to sit to avoid reflected

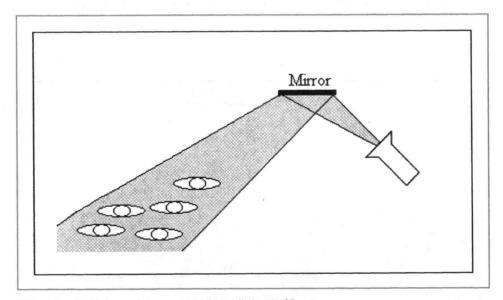

FIG. 4-9 *The Elaborated Answer to Light Prediction Problem 2*

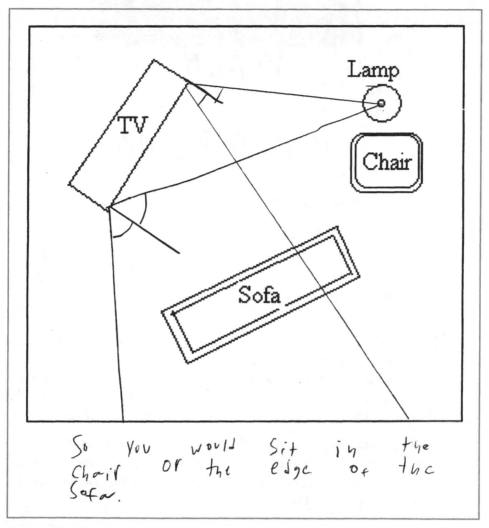

So you would sit in the chair or the edge of the sofa.

FIG. 4-10 *Elliott's Answer to Light Prediction Problem 3*

light on the TV from the lamp. Elliott's answer to the problem is presented in Figure 4-10. His answer demonstrates a clear understanding of the law of reflection and his ability to apply his knowledge to a new situation. By the end of this 5E sequence, most students had learned the law of reflection and could use it to make predictions and solve light problems. My use of seamless assessment helped me determine what they understood and decide how to challenge them as we progressed in the unit.

Vignette

Will It Float?

■ *Kelly Turnbough*

Unit Notes

Grade: 8

Learning Goals: Students will learn successively more sophisticated explanations of density: (1) density is a qualitative comparison of whether an object will float or sink in water, (2) density is a comparison of the mass of an object with its volume, (3) density is the mass-to-volume ratio of a substance, and (4) density is an intensive property that can be used to identify a substance.

National Science Education Standard: Content Standard: 5–8, Physical Science: A substance has characteristic properties, such as density, a boiling point, and solubility, all of which are independent of the amount of the sample. (National Research Council 1996, 154)

Assessment Strategies:

Engage: prediction/observation sheet

Explore: conclusion writing

Explain: letter to the teacher

Elaborate: bowling ball discrepant event

Evaluate: memo to the boss

Vignette

Engage

"What will happen when I put this in the water?" I asked, holding a small chunk of pumice above a ten-gallon fish tank.

Most students laughed and shouted out, "It will sink."

Before placing the rock into the tank, I asked the students to open their science notebooks and draw two boxes, one labeled "prediction" and the other "observation." "Draw a picture to show what you think will happen when I place this rock in the water." After the students had finished their prediction drawings, I gently placed the rock in the tank. To the students' amazement, the rock floated.

As students completed their observation drawings, I placed additional items onto my demonstration table, including a larger chunk of pumice. The class began to murmur, "Put the big rock in the water! Put the big rock in the water!" I asked them to draw two new boxes and draw their prediction in the first box. When the larger rock floated in the water, the students showed their surprise in the form of *oooh*s and *aaah*s. I hoped this demonstration would cause students to question their preconceived idea that floating is dependent on either the size or the weight of objects.

I asked, "What was different about the two rocks?"

Enrique responded, "One was bigger."

Tyrone said, "The big one weighs more."

I listed those ideas on a flipchart. "What was the same about the rocks?"

Jasmine volunteered two observations: "Both rocks had air pockets and they both floated in water." I added these characteristics to the flipchart. Jasmine's answer suggested that we should investigate the role played by air in sinking and floating, while the boys' responses suggested we should investigate mass and volume.

Explore

I initiated the air exploration by challenging students to create a shape out of foil that would float. Using equal-sized pieces of aluminum foil, they designed pieces with high sides, low sides, no sides, and a few that were wadded-up balls. The students made predictions about what would happen when they placed their foil into water. The boatlike designs floated, but the tightly packed balls of foil sank. Students drew their observations and wrote brief descriptions of what they had seen.

Matt shared, "I saw the foil boats float and the foil balls sink."

Kyle read from his notebook, "I think the reason the boats floated better than the balls was because the boats spread out and covered more area than the balls."

Deborah, who had constructed a foil ball, said, "My foil ball sank because I saw all the air bubbles escape when I squeezed it underwater and the boat still had air in it so it floated." Her explanation was convincing and almost all the students agreed that the air in the boat or in any object played some role in helping it float in water. This discussion demonstrated that students were beginning to understand that floating and sinking was related to properties of objects, but I was unclear about how they were interpreting the role of air.

For students to explore further the role of air in floating and sinking as well as the mass-volume relationship, they needed to test solid objects of various sizes without trapped air. I provided each team with blocks of equal size but unequal mass and small tubs of water and asked them to draw and write predictions about what would happen when they performed the sink/float test. Deborah's team made predictions and then tested them by putting all the blocks in the water at once. Some sank and some floated. They tested the floaters further by squeezing them to make sure there was no trapped air inside; to their delight, the blocks still floated. I noticed that Marty held different blocks in each hand to compare their masses. I heard her tell her team, "This heavy one sank but this light one floated." Students used their science notebooks to record data, including each object's volume, and draw pictures (see Figure 4-11).

At the end of the class some of the students shared their conclusions. Deborah read from her notebook, "I thought things floated because of air. I learned that things can float even if they don't have air in them. I observed that things the same size can be heavy and sink or light and float in water." Later, as I read other notebook entries, I learned that a number of students had written similar statements. For example, Kyle wrote, "I think the size of the object might make more of a difference than air."

Explain

In order to help students understand how floating and sinking are related to an object's mass and volume, I asked each student to bring a vegetable or a fruit to class. Using the foods, we played the game "Will It Float?" After students made a prediction for each item, the student who brought the food performed the sink/float test. That same student recorded the finding on our class chart in either the "floater" or the "sinker" column.

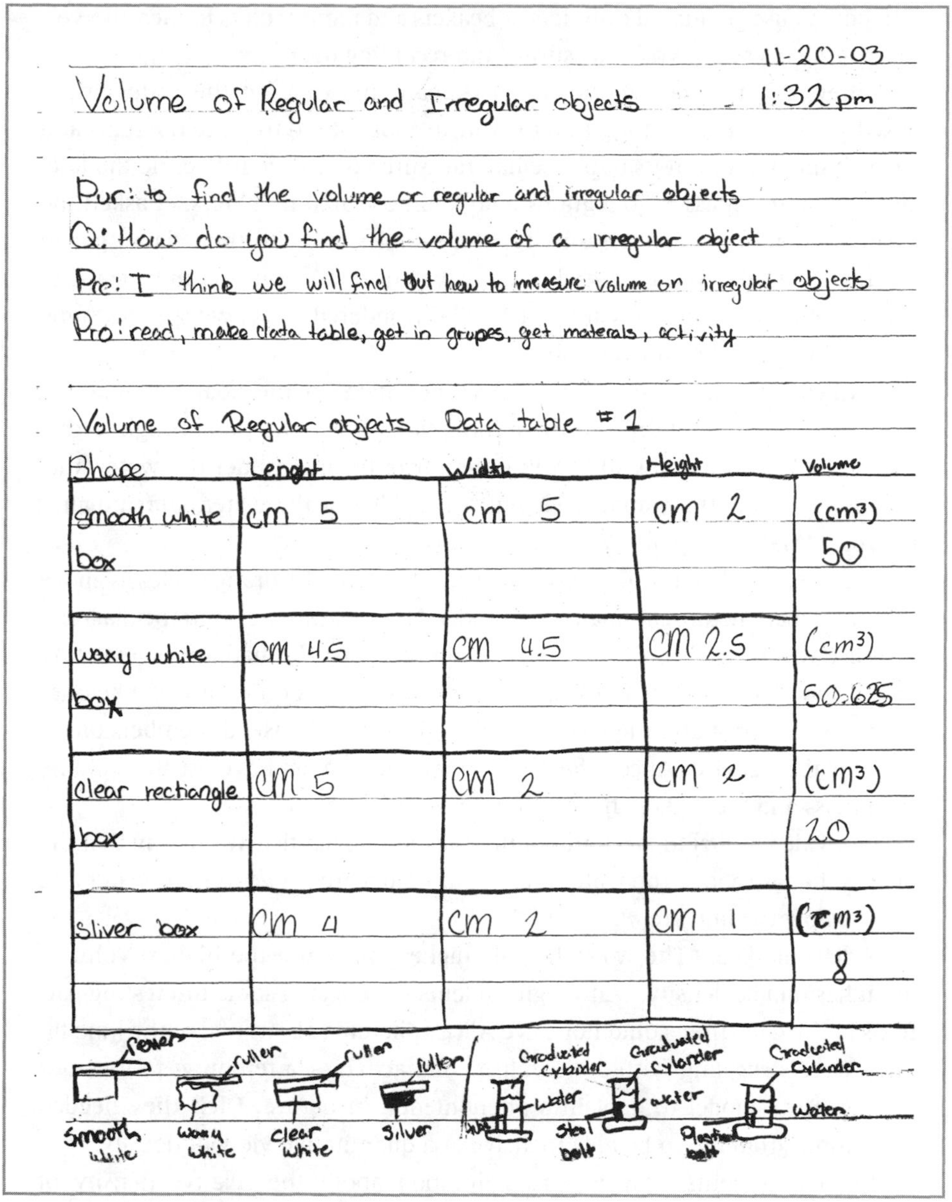

11-20-03

Volume of Regular and Irregular objects 1:32 pm

Pur: to find the volume or regular and irregular objects

Q: How do you find the volume of a irregular object

Pre: I think we will find out how to measure volume on irregular objects

Pro: read, make data table, get in grupes, get materals, activity

Volume of Regular objects Data table # 1

Shape	Leinght	Width	Height	Volume
smooth white box	cm 5	cm 5	cm 2	(cm³) 50
waxy white box	cm 4.5	cm 4.5	cm 2.5	(cm³) 50.625
clear rectangle box	cm 5	cm 2	cm 2	(cm³) 20
sliver box	cm 4	cm 2	cm 1	(cm³) 8

FIG. 4-11 *Notebook Page About Floating and Sinking*

Because students had explored density qualitatively through sinking and floating, I decided they were ready to formalize the concept of density as a quantitative property. I asked students to measure the mass and the volume of the food items. Those who did not bring in food measured the mass and the volume of a sample of water. We reviewed how to use the pan balance to measure mass and how to use graduated cylinders or beakers and metric rulers to measure volume. Students set to work measuring and recording data.

I drew a horizontal line on the blackboard and called it the water line. I asked Jeremy (who had forgotten to bring in a food item) to write the mass and the volume of the water sample he had measured on the board beside the line. Jeremy wrote, "Mass = 50.5 grams and Volume = 50.0 milliliters." I asked the other students who had measured water samples to write their data next to Jeremy's data. Deborah noticed a pattern. "The mass and volume measurements are almost equal to each other." She wondered, "Will we see this same pattern for the fruits and vegetables?"

I asked for volunteers to place their food data on the board, wondering aloud where we should record the information. Kyle, who had brought a carrot, raised his hand and said, "I think the carrot belongs under the water line, because it sank in the water." Kyle put "carrot" below the water data and after it wrote, "Mass = 75 g and Volume = 73 mls."

Enrique, who had brought an orange, said, "I think the orange belongs above the water mark because it floated," and added his measurements to the diagram.

Alice was the next to volunteer but admitted that she had a problem. "My lime belongs below the water line because it sank, but I don't know if it belongs between the carrot and the water or below the carrot." I asked members of the class for their advice. Alice reflected, "So far, the floaters have greater volume than mass and the sinkers have greater mass than volume."

I provided a label to facilitate communication. "Scientists would say that the objects that are above the water line in our model are *less dense* than water and those below the line are *more dense*."

Kyle remarked, "The water has the highest mass and the highest volume, but it has middle density." Although students were seeing some interesting patterns in the data, they could not solve Alice's dilemma about where the lime fit. They were struggling with a difficult mental activity—attempting to invent a mathematical model to explain a phenomenon in nature. I felt they needed more input from me to be able to arrive at a quantitative view of density.

To help students answer Alice's question about the relative density of objects, I handed each student a blank graph of mass versus volume. I displayed a similar graph on the overhead. Together we plotted five data points for water density. Students drew a diagonal line of best fit for these data, which I labeled "water density line." Then I asked Kyle to place his carrot data on the overhead graph. The carrot point was just below the water line. Next, Enrique plotted the point for his orange just above the density line for water. Alice plotted her point for the lime a little past the point for Kyle's carrot. Kyle exclaimed, "I see a pattern! The lime must be more dense than the carrot because its point is farther from water."

Other students nodded. After the remaining students took their turn at the overhead, I reinforced Kyle's point by stating, "How far a point is away from the density line for water indicates that object's relative density," We reached closure on this part of the lesson by ordering all of the foods from least to most dense in a list on the board.

At the end of the lesson, I asked students to write me a letter that explained one thing they had learned about density and something that confused them (students had recently practiced letter writing in language arts class). Enrique wrote:

Dear Ms. Turnbough,

I told you that the size of a thing does make it float or sink. Then you asked me if big things can't float and I think they can float—if the size of the object is about equal to the mass. But I'm still wondering if you are saying that if the mass is way bigger than the volume then the thing will sink in water but if the mass is smaller than the volume then the thing will float?

Your Humble Servant,
Enrique

Several students went further than Enrique, writing that density tells if an object will sink or float. Others wrote that the greater the density of an object, the greater its distance is from the density line for water on a graph. Only one student mentioned density in terms of the ratio of mass to volume. Alice wrote, "The density of an object is the object's mass compared to its volume."

Elaborate

The next class period, I decided to share Alice's answer with the other students, elaborating on how to calculate density. "We can find the density of any object just by dividing the mass by the volume." I performed a few sample calculations with the water data; the students quickly realized that each calculation was about one. I assigned half the class to calculate the density of the objects above the water line on our graph and half to calculate the values for those points below the line. They were impressed to find all the floaters had values less than one, while the sinkers had values greater than one. Once I was confident that the students understood and could do the calculation for density, I posed a new problem.

Pulling a large bowling ball out of my cabinet, I wondered, "Do you think a bowling ball will sink or float in water?"

"Sink!" exclaimed most of the students.

"It's a trick question," Tyrone claimed.

"Can you know if it will sink or float without putting it in water?" I asked. Deborah offered that they could use the density formula. Other students nodded. "What information do you need?" I probed. At their request, I provided the mass and volume data for the bowling ball. While individuals calculated the bowling ball's density, I drew a "Sink or Float" prediction chart on the board. "What's your prediction about the bowling ball now?" I asked. Most of the students raised a hand for float, but a few (some who had made arithmetic errors and others who refused to trust their data) stuck with sink. Next came the dramatic moment. I lowered the bowling ball into a huge tub of water. The bowling ball slipped slightly under the water and bobbed back up to the surface. The students applauded in delight.

"Do you think all bowling balls will float?" I asked next.

Jasmine, who recently had held her birthday party in a bowling alley, said, "Some of the balls are a lot heavier than that one."

"Yeah, but are they denser?" challenged Kyle. Rather than debate the issue, I passed out the "Bowling Ball Density Challenge" worksheet to the students. The sheet listed the mass and the volume for a set of five bowling balls and asked, "Will each bowling ball sink or float? Why do you think so?"

Students readily took the challenge and completed the sheet. Later that day when I examined their work, I found that most students were successful in applying the density formula, performing the calculations, and deciding which balls would sink or float. Figure 4-12 shows a page from Enrique's notebook illustrating what happened with three different bowling balls. Jasmine's answer revealed her understanding and her sense of humor: "I knew the balls were different weights. But they are the same volume. So now I know that they have different densities too. So there, Kyle!"

Evaluate

Throughout the density unit, I informally assessed students as they talked to each other and shared ideas. Now that we were at the end of the unit, I wanted a formal assessment of individual student learning. I wanted to assess students' understanding of density as a property of objects and their skills at calculating the mass-volume relationship. I handed out the following assignment:

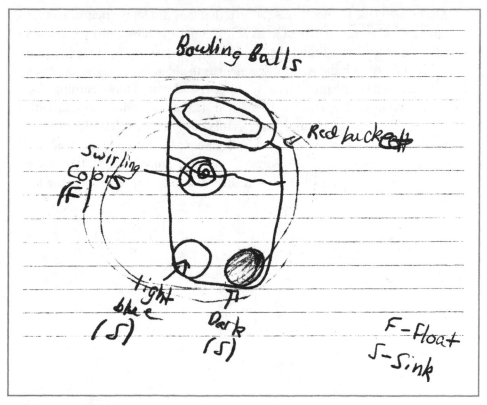

FIG. 4-12 *Enrique's Bowling Ball Drawing*

You work for the Acme Company, Floaty Toys Division. Your boss has asked you to determine which materials would be best to use to make floaty toys. Using what you know about density, decide if your material should be ordered by Acme. Write a memo to your boss that discusses the properties of your material (including density) that would make it a good or a bad choice to use for floaty toys.

I gave each student a test material and they got busy measuring, calculating, and writing. I was pleased to see students finding the mass and the volume of the materials and using their calculators to figure out density. Thinking back to the first day of the unit, I was curious to read their memos to the boss and see how their understanding about density had developed. Deborah explained to her boss, "When you decide which materials to buy for the toys, just remember 3 things: 1. if the mass is much larger than the volume, then the matter will sink. 2 if the volume is more than the mass, the matter will float. 3. if the volume and the mass are the same, the ball will float."

After reading the memos, I was satisfied that most students understood that density is a relationship between mass and volume and how density is related to sinking and floating. I found that my low-achieving and ESL students were almost as successful on the Acme assessment as the high-achieving students, as long as they could recite their memo to a tape recorder or to their learning aide instead of writing it down. In the past, I had started and ended this unit with the formula for density. This time, the formula came after much student exploration and many tentative student explanations. In the end, the students thought of density not just as a formula, but as a property of matter. More importantly, these students believed in their ability to figure science out for themselves.

Seamless Assessment in Earth and Space Science

5

■ Introduction

The vignettes in this chapter are based on 5E units of instruction carried out with primary (grades 1–3), intermediate (grades 4–5), and middle-level (grades 6–8) science students. In these units, teachers aimed to develop students' conceptual understanding about soils, rocks, volcanoes, and moon phases. Each vignette illustrates how teachers used seamless assessment to plan and inform their instruction. Table 5-1 describes the grade levels, topics, and assessment strategies used in the vignettes.

Young children are interested in everything they see around them. This interest is nurtured when teachers capitalize on observable phenomena to help students understand the complexity of their world. As students grow older, their ability to understand abstract concepts in earth and space systems grows. Teachers who challenge students' incoming conceptions of earth systems through firsthand experience help students change how they think and develop deeper understandings. That is what the units in this chapter attempt to do.

In the first vignette, Kenneth Welty challenges his second-grade students to describe the properties of soil. Their first response is that soil is "just dirt." Kenneth helps his students understand the complexity of soil by exploring its constituents. These lessons help students develop soil concepts that provide a foundation for learning about related science concepts, such as plant growth and decomposition.

Table 5-1 Earth and Space Science Vignettes by Grade Level, Topic, and Assessment Strategy

Vignette	Grade Level	Topic	Assessment Strategies
Toiling in the Soil	2	Soils	*Engage*: KWL chart *Explore*: science notebook: properties of soils observations; soil tube drawings *Explain*: science notebook: L column of KWL chart; soil equation *Elaborate/Evaluate*: "Mystery Mixture" sheet
Rock On!	4	Rocks	*Engage*: notebook: rock observations and questions *Explore*: notebook: rock group reasons *Explain*: notebook: sedimentary rock decisions *Elaborate*: notebook: rock guide research *Evaluate*: class rock cycle chart
It's Volcanic!	7	Volcanoes	*Engage*: brainstorming and preunit concept mapping *Explore*: making a claim *Explain*: volcano explanation and labeled drawing *Elaborate*: memoir *Evaluate*: postunit concept mapping
Misconceiving the Moon	8	Phases of the moon	*Engage*: moon phase explanations questionnaire *Explore*: moon notebook; moon calendar *Explain*: moon model diagram *Elaborate*: moon puzzlers *Evaluate*: moon final reflection

In the next vignette, Laura Zinszer engages her fourth-grade class in the study of geology by walking a creek bed, analyzing rocks, and asking questions. Through a series of investigations, students describe general features of rocks and use them to invent a classification system. The unit draws to a close as students speculate about rock transformations, setting the stage for understanding the rock cycle. Zinszer demonstrates the importance of thoughtful, assessment-based judgments as she makes instructional choices.

In the third vignette, Meera Sood and Cathy Carpenter report on their collective experience of teaching the topic of volcanoes to middle-level students. Most students are familiar with the surface-level descriptions of spewing lava and ash clouds, but few are aware of the plate tectonics explanation that accounts for volcanic activity. As seventh-grade students analyze data about plate movements, earthquake severity, and volcano locations, they develop a sophisticated understanding of how these catastrophic events are linked.

Students of all ages misconceive why the moon appears to change shape. Some believe moon phases are due to the earth's shadow, while others think they are caused by clouds. The final vignette in this section describes how Mark Volkmann and Sandra Abell challenged eighth-grade students' conceptions of lunar phases. Using a variety of seamless assessments, they guided students to develop their spatial reasoning abilities about heavenly bodies. This story describes how teachers can help students visualize a world taken for granted in new ways.

Each earth and space science vignette was designed and carried out at a particular grade level. However, the *National Science Education Standards*, upon which each unit was built, address grade ranges, not specific grades. We hope you will find ways to adapt these units to your local context and grade level, where age appropriate. We also encourage you to develop new units about other important earth and space science concepts for elementary and middle-level students, as detailed in the standards, within the broad categories of properties of earth materials, objects in the sky, changes in earth and sky, structure of the earth system, earth's history, and earth in the solar system (see National Research Council 1996).

Vignette

Toiling in the Soil

■ *Kenneth Welty*

Unit Notes

Grade: 2

Learning Goals: Students will understand that (1) soil is made of three components: sand, clay, and humus and (2) those components have different properties.

National Science Education Standard: Content Standard: K–4, Earth and Space Science: Soils have properties of color and texture, capacity to retain water, and ability to support the growth of many kinds of plants. (National Research Council 1996, 134)

Assessment Strategies:

> *Engage*: KWL chart
>
> *Explore*: science notebook: properties of soils observations; soil tube drawings
>
> *Explain*: science notebook: L column of KWL chart; soil equation
>
> *Elaborate/Evaluate*: "Mystery Mixture" sheet

Vignette

"Wow, I didn't think there was so much to learn about soil!" Devan exclaimed. My second-grade class was engaged in a science unit about the composition of soil. For this unit of study, I used the *Soils* teachers' guide in the Science and Technology for Children (STC) curriculum (National Science Resources Center 2002). The unit included many concepts about soils for students to learn and culminated in an experiment of growing plants in different types of soils. This vignette focuses on one of the introductory concepts in the unit: "Sand, clay, and humus are three of the basic components in soil" (National Science Resources Center 2002, 2).

Prior to starting the unit, I introduced the class to using science notebooks (Campbell and Fulton 2003) to record their questions, observations, wonderings, and procedures. I use science notebooks as a form of seamless assessment—when I read them, I notice any misconceptions or misunderstandings that my students have and I can address those problems immediately. The science notebooks thus help guide my instruction. Since this was their first time using science notebooks, I asked the second graders to include the following information in their notebooks for each science lesson:

- date, time, and page number (each page was sequentially numbered)

- question or statement

- prediction

- procedure (when appropriate)

We used our science notebooks with almost every investigation in this unit. The STC *Soils* guide had several recording sheets for student use. To help keep track of these pages, students glued them into their science notebooks near their notes for the corresponding investigation. As always, they wrote the date, time, and page number at the top of each notebook page.

Before the unit began, I had started a class compost bin. This would provide the needed humus, one of the three soil components we would study. Having our own compost bin also helped reinforce the concepts that, over time, plant material becomes part of soil and worms help with breaking down the plant material and creating humus.

Engage

I began the unit by showing the class a bag of soil. "What do you suppose this is?" I asked. As expected, several of the students replied that I was holding a bag of dirt. "Not dirt," I responded, "it's soil! As soil scientists in our new unit, you will need to use the correct terminology." I asked them to think about a time they may have worked in or with soil. "Planting seeds" was the most common response. "In your notebooks, write down two things you know about soil and two questions you have about soil." Using what they wrote in their notebooks, we then made a class KWL chart (Ogle 1986). This gave students a chance to see what others had to say and piggyback off those ideas as they brainstormed new ideas. Following are some of the ideas we listed in the K (what we know) column of the chart:

It is underground.	It is brown.
Soil helps things grow.	Soil is home for the worms.
I know soil makes things grow.	We use soil for plants.
A lot of things grow in soil.	It has chemicals in it.
Helps birds get food.	Soil is good for plants.
Soil feeds the plant.	Soil is part of the ground.
Soil has wood in it and rocks.	Soil has holes in it.

In the W column (what we want to learn), we listed the following questions:

Does soil help you?	How do plants grow in soil?
What is soil?	How do you make soil?
What is soil made of?	What grows from soil?
What is it?	What does soil do?
What lives in the soil?	Why do we need soil?
What is the little white things I see in some soil?	

This final question—What is the little white things I see in some soil?—had come from John's experience with watching his mother repot houseplants at home. We included a third column, L (what we learned) on our chart but left it blank for now. We would add to it as we proceeded through the unit.

Later, when I read their notebooks, I could tell that some students were confused by the question-asking phase. For example, Devan wrote, "Who invented soil?" Julie asked, "What year was soil made?" and Connor wrote, "What is it like in soil?" These were rather odd but interesting questions, so at the time I chose not to discuss them. Looking back, I realize they would have been worthy of discussion to find out what students wanted to know and why these questions were important to them. Many other student questions were within the scope of our study and provided a good foundation for beginning the unit.

Explore

Over the next several class periods, students explored the properties of each soil component—sand, clay, and humus. Students used four of their five senses to observe each component. They looked at the soil component with hand lenses, smelled the sample, touched the sample, and listened to what it sounded like when shaken in a cup. They also performed other tests suggested by the STC curriculum: getting each sample wet and trying to make a ball and doing a smear test to see what each component looked like on paper. Another investigation involved pouring water through each soil component and observing which one held the most water, or allowed the least amount of water to flow through into another container.

Throughout these investigations, students worked with partners so they could talk and share observations with each other. They also kept records of their observations in their science notebooks. When I asked them to draw what they saw, students were surprised. It seemed unusual to them that they could record information by using a picture instead of words. After their initial surprise, they did quite well using this form of recording. Throughout their investigations of the properties of soil components, students recorded observations in their science notebooks using words, pictures, and labels.

For their final exploration of the properties of soil components, students placed samples of each soil component in a test tube and added water. They sealed and shook the tube and recorded their observations. Students' drawings of their initial observations demonstrated that they had noticed the different colors and textures in the soil components (see Figures 5-1 and 5-2). We kept the tubes for several days to let the components settle and then made further observations. Devan drew a detailed picture of the settled components that was significantly different from his initial drawing (see Figure 5-3).

Explain

Now that we were armed with loads of observations, it was time to summarize what we had learned about sand, clay, and humus on the KWL chart. We ended up creating a new three-column chart to insert into the L column that represented student observations of each soil component (see Table 5-2). These observations demonstrated to me that students not only understood the properties of each component but had begun to make comparisons across the three types. For example, they noted: "Clay holds more water than sand or humus,"

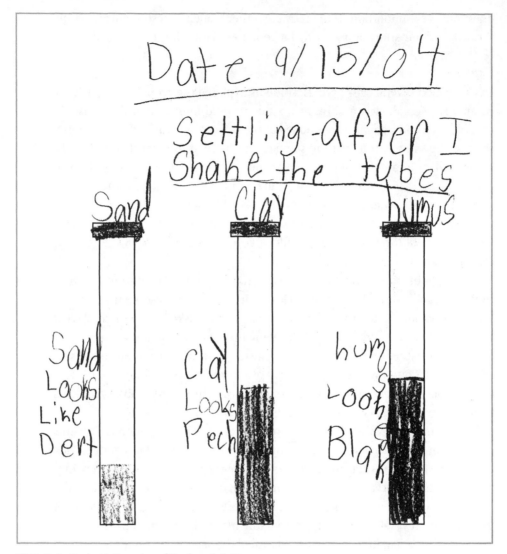

FIG. 5-1 *Student's Drawing of Explore Activity*

"Water goes through sand better than clay or humus," and "Humus is a darker color than sand or clay."

From this information, I felt that the students were ready to see how sand, clay, and humus were all components of soil. To help them understand this important concept, I took one part each of soil component—clay, humus, and sand—and asked the students what they thought would happen if I mixed the three together. Most students responded that they didn't know, so I proceeded

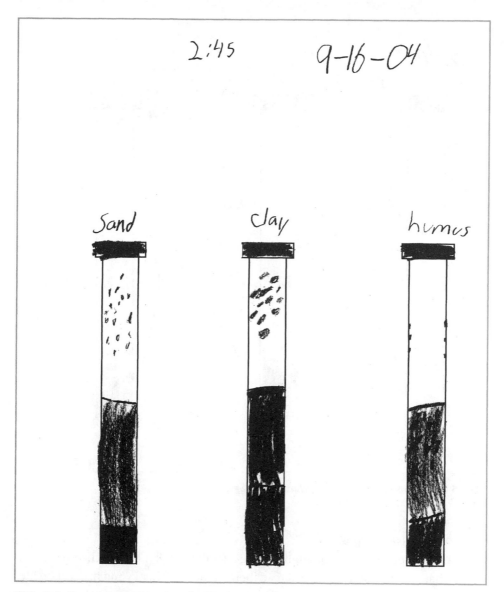

2:45 9-16-04

Sand   Clay   humus

FIG. 5-2 *Devan's Initial Drawing of Soil Components in Water*

with the demonstration. I added equal amounts of each component to a container and shook it up. I placed a container of soil next to this new mixture and asked students to compare the two. All across the room faces lit up as students realized that sand and clay and humus together make soil. They decided to write an equation in their science notebooks to represent this idea:

sand + clay + humus = soil

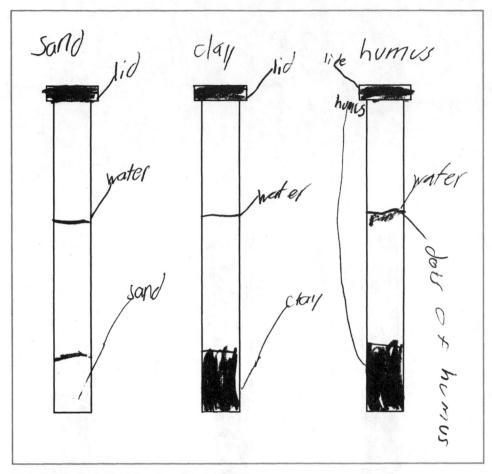

FIG. 5-3 *Devan's Final Drawing of Settled Soil Components*

Elaborate/Evaluate

Throughout this unit of study, I heard students talking to each other about what they had observed. I had read their science notebooks and had a good idea of what they were thinking. Now it was time for them to apply their developing ideas to a new problem and to deepen and broaden their understanding of soils in the process. I did this through the Mystery Mixture Challenge.

I gave each team three different mystery mixtures: mixture A had equal parts sand, clay, and humus; mixture B was two parts clay to one part sand; and mixture C was two parts clay to one part humus. I challenged students to use what they knew about properties of soil and soil components to identify the mystery components in each mixture. Each group conducted the same soil tests they had used earlier in the unit: they looked at each mixture with a hand lens,

Table 5-2 Class-Generated Properties of Sand, Clay, and Humus

Sand	Clay	Humus
Sand is hot.	Clay is like Play-Doh.	Humus is dirt.
Has little rocks in it.	Sometimes clay is found in sand.	It is a different kind of soil.
Sand is smooth.	Clay is gray.	Feels mushy.
When it gets wet, it gets mushy.	Clay is different colors.	Humus is brown.
You can build with it when it's wet.	You can build with clay.	It is another home for worms.
When sand is dry and you pick it up, it will slide out of your hand.	It is gushy and mushy.	It is dirty.
	Clay is hard.	When you shake it with water in the test tube, it floats.
When sand gets wet, it gets softer.	When clay dries up, it is hard and white.	When you shake it with water, it separates into three groups.
Is yellow.	Clay can sometimes be wet.	Humus makes a dark color in water.
When sand is dry, it's not that heavy, but water makes it heavier.	Clay is way underground.	Humus has wood in it.
	You have to dig to find clay.	When humus gets wet, it gets a darker color.
Sand sinks in water because it's like lots of tiny rocks.	When you add water it looks darker.	Humus can be grass, dead leaves, or sticks.
Sand changes the color of water little.	Clay changes the color of water.	Humus is a darker color than clay or sand.
Water goes through sand better than clay or humus.	The longer you wait and let the clay settle, almost all of the clay will go to the bottom.	
	It turns a lighter color when it dries.	
	Clay holds more water than sand or humus.	
	Clay pieces are smaller than sand.	

smelled the mixture, touched a dry sample, rolled a moist sample, conducted a smear test, and performed a settling test. They compared their observations with the outcomes of their previous explorations to help them identify what was in their new mixtures.

During this investigation, students talked about what they observed and used information from prior investigations that they had recorded in their science notebooks. At first they were not comfortable with the idea of looking back in their notebooks; some thought this might be cheating. I explained that scientists use their laboratory notebooks to help them figure out new data. By asking students to use the information from their notebooks as they evaluated the contents of the mystery mixtures, I had provided an incentive for accurate record keeping in future science investigations.

Students recorded their team's final ideas about the mystery mixtures in their individual science notebooks. I asked them to make a claim about what was in each mixture and provide evidence for their claims. By having the students identify the soil components using previously performed tests, I was able to assess how well they understood the concept that soil is made of different components. I also saw how well they could use evidence to support their explanations, one of the key features of inquiry.

Figure 5-4 presents several of the students' answers for one of their mystery mixtures in response to the final test of placing a sample of the mystery mixture in a test tube, adding water, shaking, and letting it settle. Each student was correct in the identification of that mixture and was able to cite evidence for the presumed composition. Mariah decided humus was floating on top of her mxture because she saw black dots. She also thought the mystery mixture contained sand because she saw rocks, and clay because the water was brown. Hannah decided her mystery mixture contained humus because it was black and sand because she saw rocks. Connor saw black dots (humus) and rocks (sand). Devan decorated his paper with question marks all over. He wrote that he had observed little rocks (sand) and decided the mixture also had humus because of black dots, and clay because it was the color of clay. Donetta saw black dots and little rocks in the settled mixture, black stuff floating on top, and rocks on the bottom, leading to her decision that the mystery mixture was made of sand and humus. Madison explained that she saw sand because sand looks like rocks and humus because it looks like dirt.

The class was excited and engaged with their learning about soil. I believe that they developed a deeper understanding of the concepts we were studying because of their involvement. At the end of the unit, I felt confident that they understood that soil is made of three separate parts and that each component

What Is My Mystery Mixture

What is your mystery mixture?
humus clay and sand

Why do you think so?
I so humus flowt on top
beucas I so blak dots
I thek that there is sand beckas
thar roks
I no thats clay bekas it is brwn

Mariah's assessment

What Is My Mystery Mixture

What is your mystery mixture?
it is sand and humus

Why do you think so?
becasu when I look at the
mysery mixture I now it's humus
becasu it black and I now it's
sand becasu I see rook

Hannah's assessment

What Is My Mystery Mixture

What is your mystery mixture?
I tenk sand and humus

Why do you think so?
I see Black dots.
I see roks.

Connor's assessment

What Is My Mystery Mixture

What is your mystery mixture?
Sand clay & humus

Why do you think so?
I think it is sand because I see
little rocks. I think it is humus
because it has black dots and
humus has black dots. I think it
clay because it has the color of
clay.

Devan's assessment

What Is My Mystery Mixture

What is your mystery mixture?
it is sand + homis

Why do you think so?
I soe Black dots and litall Rocks
in my sitalling thar was Black
stof floting on top and rocks on
the Botom and som mor Black
sofe on the botam.

Donetta's assessment

What Is My Mystery Mixture

What is your mystery mixture?
sand and humus

Why do you think so?
it looks like it becese sand looks
like rocks and humu look like
drit

Madison's assessment

FIG. 5-4 *Selected Student Responses to Mystery Mixtures*

is different. As their teacher, I was able to assess their learning seamlessly—I knew more about what they had learned than if I had done traditional pre- and posttests. Throughout this unit of study, I had many opportunities to assess students' thinking and understanding as they made observations and applied what they knew to new situations. Using seamless assessment allowed me to restructure my teaching on the spot, thus facilitating student learning. By listening to their discussions, reading their science notebooks, and keeping track of their learning on the KWL chart, I was able to watch student learning in action.

Vignette

Rock On!

■ *Laura Zinszer*

Unit Notes

Grade: 4

Learning Goals: Students will understand that (1) rocks have different properties that can help distinguish them and (2) sedimentary, igneous, and metamorphic rock are related through the rock cycle.

National Science Education Standard: Content Standard: K–4, Earth and Space Science: Earth materials are solid rocks and soils, water, and the gases of the atmosphere. The varied materials have different physical and chemical properties. (National Research Council 1996, 134)

Assessment Strategies:

   Engage: notebook: rock observations and questions

   Explore: notebook: rock group reasons

   Explain: notebook: sedimentary rock decisions

   Elaborate: notebook: rock guide research

   Evaluate: class rock cycle chart

Vignette

"Ms. Zinszer, I've found something shiny inside my rock!" Gabrielle couldn't contain her excitement as she stood poised over her rock with hand lens and notebook. The fourth graders in her group gathered around her and agreed there was something special about that rock. Questions began to emanate from the students. "What is that shiny stuff inside the rock?" "How did it get stuck inside, anyway?" "Where did all of these rocks come from?"

Engage

The fourth graders had just begun the rock unit in science, the first unit of the school year. Each year, I begin the unit a little differently. This year, I took the students on a short hike along the creek bed behind the school to search for rocks to study. This was before they had any information about how rocks form, different types of rocks, or the material that can make up different rocks. My primary goal for this introductory lesson was for the students to become engaged with the study of rocks by having an opportunity to find a rock, make observations, and ask questions. I hoped that the students would make observations about their rocks and that the questions would develop naturally from their observations.

As the students collected their rocks, they recorded observations in their science notebooks. Mackenzie pointed out, "My rock has different colored lines that stack up on top of each other." Sara wrote, "My rock has a fossil sticking all the way through it." Some students drew their rock and used arrows with labels to identify their observations. Other students seemed more comfortable writing a narrative description. I encouraged students to use any method that they wanted to record their ideas. While they were making and recording their observations, some discussion ensued about how to figure out which observations were important enough to write about.

Once the observations in the field were finished, the students headed back into the school building. When we returned to the classroom, the budding geologists settled on the floor in front of a large sheet of chart paper with a large rock drawn on it. It was time for the students to formulate different questions about the rocks from the observations they had been making. Having the students develop questions from their rock observations provided an excellent assessment for this engagement activity. First, I asked students to write in their science notebook at least two questions based on their rock observations. After a few minutes, I asked the students to pick a question they wanted to share with the class.

Lashawntay started the sharing with his question: "Where did all of the rocks in the creek come from?" Quickly others joined in. "Why do some rocks have fossils?" "What do the lines in the rock mean?" "Do rocks ever change?" "Where do rocks come from?" "How did the rocks get crystals in them?" "Are rocks and minerals the same thing?" "Are rocks all the same or different from each other?" "What is that shiny stuff inside my rock?" "Do volcanoes help us get rocks?" As the students presented, they recorded their question on the chart paper inside the outline of the rock. While each presenter shared, others could add the question to their own notebook list if they thought it was important to think about.

From my perspective, it was essential that the students' questions covered all aspects for understanding rocks, because we would use these questions to lead the lessons throughout the unit. I was surprised by the complex thinking and in-depth reflection the children used when developing their questions. I found I did not need to insert additional questions because the students had done such a thorough job.

Explore

The next lesson involved exploring a set of rocks that I had prepared for the students. As I brought in the rock sets and arranged them on tables for each group, the students chatted excitedly. Each pair of students had a hand lens, a science notebook, and a flexible centimeter ruler. The students were excited to research our question for the day, Are all rocks the same, or are some different from each other? My goal was for the students to explore the rock sets and determine if there might be ways to use their observations to group different rocks together.

I asked the students to look at the rocks carefully. "What observations can you make about the different rocks?" Each pair of students had about ten minutes to look at the rocks in their set and draw, label, and list observations about the different rocks. They identified each rock in the set by a different letter. Once the students had explored the rocks in their set, I asked them, "Can you and your partner organize your rocks into five different groups based on their traits?"

Most teams first tried to group the rocks by color. They quickly figured out that there were too many colors to make only five groups. Mackenzie and her partner observed that several of the rocks had layers like the rock she had found in the creek. They wondered if rocks with layers that were straight belonged in a different group from rocks with bent layers. Brian and Tim tried to make a group of rocks with crystals, but they had a problem because some of the rocks with crystals had very thin layers and other rocks with crystals had no layers. Lizzie thought about this and suggested, "Maybe the groups could have more than one trait, like crystals *and* layers." The rest of the class thought this was a great idea and most of the teams began to create groups of rocks with more than one trait. Some of the groups were layers and fossils, holes, holes and crystals, only crystals, crystals and layers, thick layers, thin layers, and layers with "other stuff."

Next I challenged the students to find the best way to divide the rocks in their set into only three groups. They considered my questions as I interacted with the teams: "Which traits did you use to make your three groups? What rocks fit into these groups? Why did you select these traits to divide your rock set? Were there any rocks that didn't fit into any of your groups?" To reach closure on this activity, I instructed students to write in their individual science notebooks, "Describe your best three groups and list the rocks that fit into those groups." Students also shared their ideas with the rest of the class. I found the reasoning required for the selection of the three groups a valuable class activity; it formed a foundation for discussing the ways that scientists classify rocks.

Explain and Elaborate

"Ms. Zinszer, I thought we were studying rocks. Why is all this dirt all over our tables?" asked Tierra.

"Actually, that's not dirt but different types of sediment. Today we are going to use this sediment to see if we can answer some of our questions about how rocks form and the types of rocks found here in Missouri." Each table had separate containers of sand, silt, humus, gravel, and clay. I handed each pair of students a clear plastic cup and a plastic spoon and invited them to scoop a sample of different sediments into their cups in any order they chose. Once they were finished, they drew an illustration of their cup and labeled the samples of sediment.

"Imagine these sediments were found in the Missouri River. What do you think would happen if we added about one hundred milliliters of water to this cup and stirred it up like the river would?" I asked.

Tyler announced, "All the stuff will just get mixed up in the cup and you won't be able to tell what is what."

"No, I don't think so," Lashawntay replied. He was sure that the "heavier stuff" like the gravel would go to the bottom and the "lighter stuff" would go to the top. After a short discussion, I asked students to write their predictions in their science notebook.

Each team then added one hundred milliliters of water to their cup and began to stir their sediments to simulate the movement of the river. "Look, it's getting all muddy. So that's why they call it the Big Muddy," Shane announced, referring to the Missouri River.

"Hey, the gravel isn't on the bottom; the sand is underneath the gravel," Lashawntay observed.

Tim commented, "My cup is forming layers. These layers look like the ones we saw yesterday in our rocks. So that's how they got in there!"

I used Tim's connection to our previous lessons to help others put the ideas together. "Do you think rock could form layers like this in real life?" I asked. The students remembered several of the rocks from their set that had flat layers just like the sample in their cup. I added, "We called these different materials in the cup *sediments*. The rocks that form from these different sediments are called . . ." Already several students were mouthing the answer, "Sedimentary rocks!"

As the students drew an illustration of the changes in their cup and labeled the layers, I asked them to think about how this activity related to the rocks we had seen the previous day. I brought the rock sets back out. The students selected which rocks fit into our new category of sedimentary rocks. The students agreed that the rocks in this group would have layers that were flat and made of different kinds of sediments. Each student listed the letters of the rocks that matched his or her best ideas about what a sedimentary rock would look like. The students applied what they had learned about sedimentary rocks to selecting the rocks. I was able to see each student's understanding of sedimentary by looking at the selections he or she made.

During the next lesson, students elaborated on their understanding of sedimentary rocks. "Today, we are going to examine the decisions we made about which rocks were sedimentary," I announced. I handed each team a rock guide that included descriptions of different types of rocks, a hand lens, and their rock sets. The partners used the guide to identify and name the sedimentary rocks in their set and find out how they formed. The students easily confirmed that limestone, shale, conglomerate, and sandstone were sedimentary rocks. They were surprised that mozarkite, the state rock of Missouri, was considered a sedimentary rock, because its layers were not so clearly distinguishable.

Many of the students made connections between the fossils found in limestone and the formation of the rock in shallow seas. "Ms. Zinszer, was Missouri really an ocean years and years ago?" This question led into an excellent discussion of fossil formation and the geologic history of Missouri. The student notebooks provided a means to assess their research on the different sedimentary rocks and their understanding of how each sedimentary rock formed. Mariana, for example, divided a page of her notebook into six squares. In each of five squares she drew one of the sedimentary rock samples (shale, limestone, conglomerate, sandstone, and mozarkite) and described it. In the sixth square, she summarized sedimentary rock formation: "Sedimentary rocks form in layers under oceans, rivers, or in deserts. The rocks are mostly soft and easy to break. They are from clay, mud, sand, and gravel."

The unit continued with similar explain and elaborate sequences for igneous and metamorphic rocks. In the interest of space, I have chosen to tell only about the sedimentary rock lessons.

Evaluate

Once our class had finished *Explain* and *Elaborate* lessons for sedimentary, igneous, and metamorphic rocks, it was time to see how much the students had learned about rocks and rock formation. I planned my *Evaluate* activity to be a brief review and summative evaluation of students' understanding of different rock types. However, the students pushed the lesson even further. Here's what happened.

I wrote at the overhead projector while students worked in their science notebooks. First, we drew a circle and labeled it "Sedimentary Rocks." "What different things have we learned about sedimentary rock in this unit?" I asked the class.

"Well," replied Brian, "we know that sedimentary rocks come from different sediments." We listed the sediments in the circle. I asked the class where we might find sedimentary rocks.

"Under an ocean." "Under the desert." "In a river bed." "Just under the ground." The students decided that lots of different kinds of rocks could be found under the ground, so we should not include that idea in the sedimentary circle.

"How about important traits or characteristics we observed about the sedimentary rocks?" I asked.

"They are mostly flat with layers and can be lots of different colors," Tyler replied. Thus, we continued to fill in our circle for sedimentary rocks.

Once this was finished, we made a second circle for igneous rocks. Once this was filled in, we created a third circle for metamorphic rocks, thus producing a three-circle model of rock types (see Figure 5-5). Suddenly, Lashawntay wondered, "Ms. Zinszer, doesn't the igneous rock turn into the sedimentary rock when it is smashed and ground up by wind or water?" I had not planned on introducing the connections among the different rock groups yet, but the class was taking the lesson there, so I went along for the ride.

I turned Lashawntay's question to the class: "What do you think, Class—can we make a connection arrow from the igneous rock to the sedimentary rock?" The students nodded. "Are there any other connections we could make among the rock groups?"

Tim quickly observed, "How about igneous going to metamorphic? If it gets squeezed under heat and pressure, igneous will turn into metamorphic."

"Excellent connection, Tim." More kids raised their hands, eager to get their own ideas into the discussion and on the rock chart. I called on Brittany in the back of the class.

"I was thinking that sedimentary could also go back to igneous if it was pushed down and melted inside the earth."

"Hey, so could metamorphic," Tierra added.

"Yes!" Tyler jumped in. "And how about sedimentary turning to metamorphic also, if it gets squeezed in the earth?"

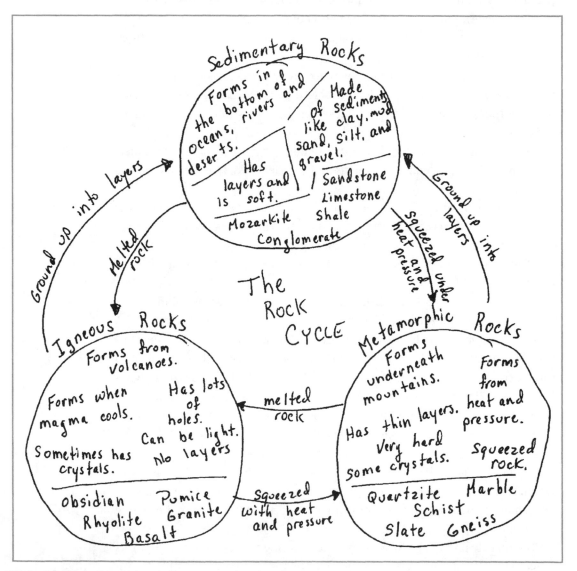

FIG. 5-5 *Teacher's Drawing of Class-Generated Rock Cycle*

We were creating arrows back and forth between the rock circles when Lawshawntay announced, "I think every kind of rock can go into every other kind of rock, like in a circle." The class stopped and thought about this.

Brian made another connection. "We have cycles of things in social studies. It's like one of those cycles that goes around and around and keeps going and going forever!"

Shanice was practically out of her chair as she exclaimed, "It's a cycle, a rock cycle!" We decided this would be an excellent name for the connections among our circles. As the discussion in class progressed, I heard several students say, "I get it; I get how this works!" When we had made all of the connections possible, the students proudly labeled our class diagram "The Rock Cycle" (see Figure 5-5).

"What have you learned about the connections between each type of rock?" I asked. As the students wrote their ideas in their notebooks, I reflected on this wonderful, serendipitous lesson. What had begun as a simple way to review the three different types of rocks had rapidly evolved into one of the most meaningful learning experiences of my science teaching. The seamless assessment activities throughout the unit led us naturally into each next phase. My students "invented" their ideas about rock types and the rock cycle, and their ideas fit with those of geologists. I learned that paying attention to student ideas and allowing those ideas to help guide the direction of instruction had a huge payoff in the end.

Vignette

It's Volcanic!

■ *Meera Sood and Cathy Carpenter*

Unit Notes

Grade: 7

Learning Goals: Students will understand that (1) volcanoes often occur where oceanic and continental plates collide and (2) earthquakes are also related to plate tectonic activity.

National Science Education Standard: Content Standard: 5–8, Earth and Space Science: Lithospheric plates on the scales of continents and oceans constantly move at rates of centimeters per year in response to movements in the mantle. Major geological events, such as earthquakes, volcanic eruptions, and mountain building, result from these plate motions. (National Research Council 1996, 160)

Assessment Strategies:

  Engage: brainstorming and preunit concept mapping

  Explore: making a claim

  Explain: volcano explanation and labeled drawing

  Elaborate: memoir

  Evaluate: postunit concept mapping

Vignette

Although we work in different buildings in our district, we share a love for teaching middle-level children. This vignette represents our combined experiences of teaching plate tectonics to seventh graders. We draw upon the

Catastrophic Events unit from the Science and Technology Concepts for Middle Schools curriculum (National Science Resources Center 2000). This sequence focuses on volcanoes as evidence for plate tectonics—the organizing theory for the earth sciences.

Engage

We began the volcanoes unit by engaging students in a staged reading of a play called *The Wrath of Vesuvius* (see Conklin 2000). In small groups, students read aloud, taking on one or more of the eight roles in the play. Vesuvius was primarily an ash-and-pyroclastic-flow type of eruption. From reading the play, students were impressed that Pliny the Younger, in AD 79, was the first person ever to describe such a flow. The students were also surprised to find no description of a red-hot lava flow, since many students think that is the only kind of volcanic eruption. Reading the play ignited students' interest in volcanoes and motivated them to learn more.

To follow up the reading of the play, we next engaged students in a brainstorming activity. "What are some words associated with volcanic eruptions?" The seventh graders eagerly shouted out answers: *magma, lava, volcanic rock, ash, gases, eruptions, plates, pressure, earthquakes, mantle,* and *crust*. Working individually, students constructed concept maps using these words and others and described the connections among the terms. Figure 5-6 shows Emily's initial concept map. She placed the terms on the map, but did not note any connections among them.

When students completed their concept maps, they wrote an answer to the question Why do you think volcanoes erupt? Their answers included "Volcanoes erupt because of pressure built up from moving plates underground," "Earthquakes make volcanoes," "Volcanoes erupt from the pressure of magma building up below it," "Hot air beneath the lava causes the volcano to explode," "Shifting underground plates cause the magma to jump up," and "Volcanoes happen at the edge of plates." These answers indicated that students saw a connection between volcanoes, earthquakes, plates, and magma. However, concept maps like Emily's indicated that these connections were not based on an understanding of the underlying theory: plate tectonics. We used their responses to help us decide what to teach. In addition to our initial question (Why do volcanoes erupt?), we focused this beginning 5E cycle on three other questions: Where are volcanoes formed? Why are they formed there? and What is the relationship between volcanoes and other catastrophic events? We believed these questions would help our students understand volcanoes as evidence of the larger theoretical system called plate tectonics.

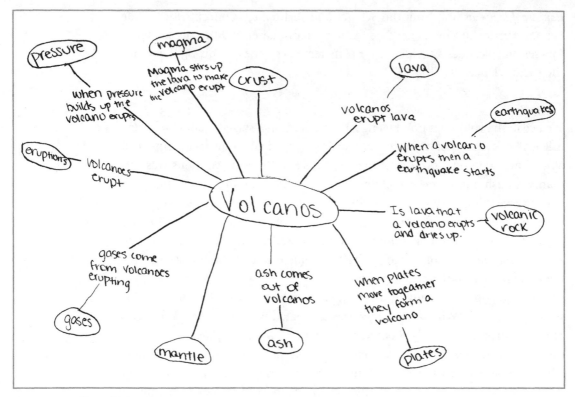

FIG. 5-6 *Emily's Initial Concept Map*

Explore

To help students arrange the data they would collect throughout the volcano study, we instructed them on how to use a graphic organizer (see Zike 2001). Students made a three-column table with the following labels: "Where Volcanoes Occur," "The Different Types of Volcanoes," and "Material That Erupts from Volcanoes." Although our first 5E cycle would focus only on the location and formation of volcanoes in relation to oceanic and continental plates, in later cycles we would need the other columns of the graphic organizer.

We also thought these Midwestern students would need an immediate connection to volcanoes throughout our study to supplement the more abstract ideas we would be addressing. To achieve this personal connection, we began reading *The Volcano Disaster*, by Peg Kehret (1998). This adolescent novel is a suspense-filled adventure story with authentic historical details about the 1980 eruption of Mount Saint Helens. We incorporated a number of reading activities, including book groups, silent reading, and read-alouds, throughout the volcano unit. Students completed book summaries and used the book to add data to their graphic organizers. This novel reading continued throughout the volcano study.

With these organizing tools in place, we were ready to explore the location of volcanoes. In order to help students understand why volcanoes erupt in terms of the plate tectonic explanation, we decided to first explore where volcanoes are formed. We gave each pair of students a map of the world with latitude and longitude intervals marked on the edges and a list of volcanoes that had erupted over the past ten years. Using the latitude and longitude information, students placed an X on the map for each volcano on the list.

When they were finished, we asked, "What claims can you make about where volcanoes are located?"

Ted said, "Volcanoes happen near coastlines."

Emily claimed, "Most volcanoes are located on land and not in the water."

Yet Kiren noticed, "Not all volcanoes occur near coastlines—the Hawaiian Islands are located in the middle of the Pacific." This claim-making activity helped us plan how we would address the next question: Why are volcanoes formed there?

Explain

In order to help students connect volcano locations to plate boundaries, we introduced the idea of continental drift. The idea that continents were once a single land mass that split apart was first suggested by Alfred Wegener in 1915 (Gribbin 2002). Wegener used the idea that continental borders fit together like a jigsaw puzzle as one source of evidence for his theory of continental drift (additionally, he used fossil evidence and rock strata evidence). We used this idea as the basis for an activity. We asked students to cut out paper continents and try to reassemble the continents as a single land mass. We told the story of Alfred Wegener's development of the theory of continental drift and used the puzzle activity to demonstrate how the continents had broken into pieces and drifted to their present positions. Our students agreed that the pieces seemed to fit together, but they had trouble accepting the idea that the continents had drifted. Our students questioned this idea, just as Wegener's critics had questioned how continents of solid rock could drift.

We asked students to suspend judgment while we examined more evidence. We gave each pair of students a map of the earth showing the major plates, their direction of movement, indicated by arrows, and their approximate speed. The students studied these maps. We asked them to find and circle areas where it looked like there might be collisions between plates (where arrows drawn on one plate were perpendicular to or pointed at arrows of the adjoining plate).

Next we reintroduced the volcano maps from the previous lesson. Students compared their volcano maps with the plate maps and readily realized that most volcanoes are located near the places where oceanic plates collide into continental plates. Kaya observed, "Continental plates sliding past each other do not seem to produce volcanoes."

Always the astute observer, Ben brought up Kiren's previous comment about Hawaii. "But that still doesn't explain what's happening in Hawaii—there are no continental plates there!"

Students had found evidence that not all volcanoes are formed in the same way. "We are going to ignore Hawaii right now and examine how those other volcanoes are formed," we stated.

In order to help students visualize what happens when oceanic and continental plates collide, we examined a diagram of a subduction zone pictured in the *Catastrophic Events* student book (see National Science Resources Center 2000, 175). We asked students, in groups of four, to choose one volcano from their volcano map located where an oceanic plate and a continental plate collide. Then we asked each student to use the textbook diagram to explain the formation of his or her group's volcano. Students made drawings of plates, labeled their plates, and wrote about the volcano formation using the idea of subduction.

After working on their group's explanations, students reassembled into new groups with three students from different groups and compared their explanations with each other. We noticed that in these new groups, students listened carefully and compared the presenter's explanation and diagram with their own. Back in their original groups, they refined their diagrams and submitted a final group explanation. Their diagrams and discussion demonstrated that they understood why volcanoes formed where plates met and led into the final question of the unit.

Elaborate

During earlier class discussions, students had noticed that mountains, like volcanoes, occurred in areas where oceanic plates were moving into continental plates. They had also conjectured that earthquakes occurred in the same areas, but they had no evidence to support their thinking. To address the last question—What is the relationship between volcanoes and other catastrophic events?—we asked students to list events that they thought were related to volcanoes. Their list included earthquakes, mountain building, lava flows, explosions, and raining ash.

Rather than deal with all of these ideas simultaneously, we decided to focus on earthquakes. We thought that asking students to locate earthquakes on plate maps would help them develop a deeper understanding of how volcanoes, earthquakes, and plate tectonics are related. The empirical data indicated that earthquakes did occur around volcanic areas, where oceanic and continental plates collide. Several students, including Neela, noticed that earthquakes also occurred in some areas where volcanoes did not. "See, these plates have arrows going opposite each other, not into each other." These students helped others see that some earthquakes are located at boundaries where plates slide past each other, such as at the San Andreas fault.

As we approached the end of the unit, we asked students to reflect on our lessons and write a memoir: "Think back about what you know about volcanoes, earthquakes, and plate tectonics. What are three claims you can make? What is your evidence for each claim? What is one question you still have?" Student memoirs used writing and drawing to explain where and how earthquakes and volcanoes formed. Most students talked about colliding plates in their explanations, some included the term *subduction*, and a few students who had caught on to Kaya's observation of sliding plates described why earthquakes occurred in some places where volcanoes did not. Ben and Kiren still had a burning question, "What's up with Hawaii?" This and other student questions like "How do tsunamis fit in?" and "Why don't all volcanoes have lava?" would lead us into new 5E cycles in the volcanoes unit.

Evaluate

During this last phase of instruction, we wanted to help students see what they had learned in our short unit. Because we believe that concept maps are a powerful way to assess what students understand, we decided to return to the concept mapping activity used at the start of the unit. We asked students individually to draw a new concept map, using the terms from the initial concept mapping task as well as some new terms: *evidence, plate tectonics, subduction, mountains, colliding plates, sliding plates,* and *continental drift.* We evaluated the maps for the number of terms used, the accuracy of the connections, and complexity.

Through their maps, students demonstrated their understanding of the relationship among major ideas of the unit: continental drift, plate tectonics, and subduction zones (see Emily's final concept map in Figure 5-7). In addition, many used volcanoes and earthquakes as evidence for the theory of plate tectonics. We ended this part of the "Catastrophic Events" unit by showing

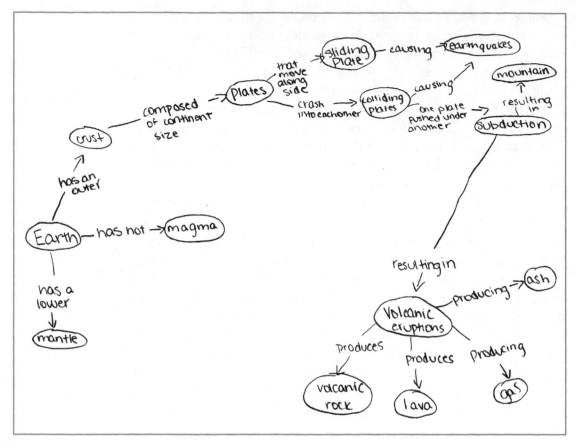

FIG. 5-7 *Emily's Final Concept Map*

students their first concept map and asking them to compare it with their second map. "What do you notice about your learning?" we asked. Students were amazed to see their growth and were proud of their understanding of some new scientific terms and ideas. As teachers, we were satisfied that we had accomplished the learning goals we had set. Throughout the unit, our seamless assessment had helped us see what students understood and paved the way for subsequent lessons.

Misconceiving the Moon

■ *Mark Volkmann and Sandra Abell*

Unit Notes

Grade: 8

Learning Goals: Students will understand that (1) we see the moon because it reflects light from the sun; (2) the phases of the moon are regular and predictable; and (3) we see the different phases because the moon moves through its orbit around the earth, showing us different portions of the lighted surface.

National Science Education Standard: Content Standard: 5–8, Earth and Space Science: Most objects in the solar system are in regular and predictable motion. Those motions explain such phenomena as the day, the year, phases of the moon, and eclipses. (National Research Council 1996, 160)

Assessment Strategies:

    Engage: moon phase explanations questionnaire

    Explore: moon notebook; moon calendar

    Explain: moon model diagram

    Elaborate: moon puzzlers

    Evaluate: moon phase final reflection

Note: A previous version of this vignette appeared in Volkmann and Abell (2003).

Vignette

Engage

"I think the moon has different shapes because the earth casts a shadow on it," wrote Matt. "I think it has different shapes because clouds cover it," Anne claimed. "I think it has different shapes because the sun makes a shadow on the moon," Mimi remarked. Matt, Anne, and Mimi were eighth-grade students who had written these explanations in response to the question, Why do you think the moon's shape changes? A couple of students wrote an answer closer to the scientific explanation—that the relative positions of the earth, the moon, and the sun create the phases. This variety of explanations is typical for middle school students who are struggling to understand the relationships among the sun, the earth, and the moon. Because students demonstrated many of the common misconceptions found in the research (Driver et al. 1994), we decided that further exploration of the moon was needed.

Explore

"You have a lot of interesting ideas about the moon. For the next several weeks, you are going to be astronomers, observing the moon's patterns and working on explanations for your observations." We passed out notebooks and asked students to record their moon observations over the coming weeks. "What information will be important to include?" we wondered. The students generated a list: time, atmospheric conditions, location of the observer, direction, and what the moon looked like. The next day students returned with some observations that surprised them, such as "The moon was only half lit," "I saw stars, but not the moon," "I saw the moon at 7 P.M., but not two hours later," and "I saw the moon right after school!" We recorded their observations on the class moon calendar by drawing pictures and listing the times when they had spotted the moon.

After several days of observing and recording, students started to raise some important questions about their observations. "I haven't seen the moon yet," complained Avery.

"Where are you looking?" Aiden asked. This question led to a quick lesson on finding cardinal directions using compasses. We issued small compasses to the students so that they could make more accurate recordings of their moon sightings. Students soon realized that when sightings were made, the observer was always facing the southern sky. Other students wanted to be more accurate about the height of the moon in the sky. We took some time to make astrolabes from protractors, straws, string, and washers (see National Research Council 2000, 49; Smith 2003) and practiced using them to measure the altitude of objects above the horizon. Using their astrolabes during subsequent observations, students added information about moon altitude to their notebooks. We also included these data on the class moon calendar.

Once students had made observations over one week, we asked them to work in small groups to analyze the data. "What patterns do you notice? What predictions can you make?" The students noticed that the lighted part of the moon was getting larger. Some sightings were occurring later into the night. Students could no longer find the moon right after school. Their predictions extended the patterns they were seeing: "I think it's gonna be full soon" and "I bet you'll be able to see it in the morning."

In order to test their predictions and collect further evidence about the moon's patterns, we decided that further explorations were in order. "We've been pretty random about observing the moon. Maybe we could get more systematic about our observations," we said. We asked each group to design an investigation that would help them think more about the moon. Groups planned to

- look for the moon each night at the same time for five days in a row

- look for the moon each hour over the course of one night (This group had a sleepover and recorded their observations through the wee hours!)

- look in the newspaper for four days to see what information it provided about the moon

- search for the moon on the Internet to see what it looked like on the days we did not see it

These new investigations provided new data for discussion. The patterns in the moon's shape were becoming clear. We provided terms (*first quarter, third quarter, waxing, waning*) to label the different views of the moon. Yet the students still had many questions: "Why does it get bigger?" "Will it get smaller?" "Isn't the moon always there? If so, why can't I see it sometimes?" Armed with patterns, predictions, and questions, we entered into the next phase of the moon study.

Explain

To begin formalizing an explanation for the patterns we had seen, we asked students to tell us what they knew about the movements of the earth, the moon, and the sun. Although they were confused about the words *revolve* and *rotate*, students knew that the earth orbits the sun and the moon orbits the earth. "We're going to use that information to build a model that might help us explain what we've been seeing," we said. We turned the overhead projector toward the class to represent the sun and passed out Ping-Pong balls to each small group to represent the moon. We also gave each group a meter stick.

"This stick will represent gravity. Tape your moon to the end." The last part of the model was one of the students in the group, whose head became the earth. That student stood up and placed the free end of the meter stick at his chin; at a slight angle above his head was the moon. The model was ready![1]

The room lights went out and the overhead projector's light glowed. The "moons" revolved slowly around the "earths," demonstrating the phases of the moon. Group members took turns observing the moon by standing behind the earth. "Oh my gosh, it's a crescent!" "I see the first quarter." "Wow, the moon is all dark now." To check for understanding, we asked all of the earths to position their sticks to make a full moon. "Now make a first quarter. Now a third quarter. Can you position the moon so it's not lit at all? That's called the new moon."

After the students observed and discussed the models, we asked them to individually replicate the moon phases in their notebook by drawing a two-dimensional representation of the moon model as if they were looking at it from space. Students drew the relative positions of the earth, the sun, and the moon in the new moon, first quarter, full moon, and third quarter phases. We asked them to explain why the moon appeared different at these times. In looking at their notebooks, we noticed that some students had trouble moving from the three-dimensional to the two-dimensional. However, most students explained that sunlight reflects off the moon to the earth and that the moon phases are due to the relative positions of the earth, the sun, and the moon. Many described the differences between certain phases as examples.

Although students were relatively secure in their understanding of moon phases, they were still pondering the patterns they had noticed in the moon's movement. To help them think more deeply about the moon's apparent movement, we brought the stick-and-ball model back out and demonstrated with one student "earth." The student positioned herself at the full-moon stage, and we added a twist to the model. Instead of having the moon orbit the earth, we asked the earth to rotate. Another student held the moon in place while the earth rotated. "Earth, your eyeball is where we live, in the Northern Hemisphere." We asked the earth to stop rotating at what she thought would be noon (facing the sun). "Can you see the full moon?" All of the students realized she could not. "Stop when you can see the moon out of the corner of your eye without turning your head." She stopped.

Paul and Felix blurted out, "That's sunrise!"

The earth continued to rotate. "Stop when you can no longer see the moon."

When she stopped, Paul and Felix, along with Avery and Aiden, exclaimed, "That's sunset!"

We turned the activity over to the small groups and asked them as a team to explore the moon model further and record the approximate times that the moon would rise and set for each of the four phases on their two-dimensional moon models. While they worked, we heard comments such as "Now I get why we couldn't see the moon at 8 P.M. for a while" and "Hey, we should be able to see the waning crescent tomorrow morning." The moon model had helped them visualize a complicated set of relationships and movements.

Elaborate

In the *Elaborate* phase of the instructional sequence, we wanted to see how well the students could apply their current understanding to new problems. It was time to introduce Moon Puzzlers. The Moon Puzzler (Abell, George, and Martini 2002) is reminiscent of thought experiments conducted by great scientists such as Galileo and Einstein. Moon Puzzlers—essentially thought-provoking questions—would help students think more deeply about the moon and test out their moon model. Students responded to these thought experiments in small groups. Here are some of the questions we asked and the answers we expected:

1. Here is what the newspaper reported for a week about the rise and set times of the moon. What phase do you think the moon is in? How can you tell? (Students could use the moon diagrams in their notebooks for assistance. The rise and set times should correspond to their notes.)

2. Why don't lunar and solar eclipses occur during each lunar cycle? (Lunar eclipses occur when the earth blocks sunlight from reaching the moon [when the moon is full], and solar eclipses occur when the moon blocks sunlight from reaching the earth [when the moon is new]. Because the moon's orbit is slightly inclined with respect to the plane of the earth's orbit, we experience full moons and new moons regularly, but lunar and solar eclipses rarely.)

3. Imagine what the first-quarter and third-quarter moons would look like in Australia. (The moon's apparent shape during each phase—except full—appears the opposite to Australians because an observer in the Southern Hemisphere is facing north when viewing the moon, while an observer in the Northern Hemisphere is facing south.)

4. In Crockett Johnson's book *Harold and the Purple Crayon* (1981), a young boy named Harold gets lost and searches for his way home. Then he remembers something. "He remembered where his bedroom window was, when there

was a moon. It was always right around the moon." What do think about Harold's explanation?

(This cannot be so. The position of the moon relative to his window will change throughout one day and night and from day to day.)

We structured the Moon Puzzlers to be of varying degrees of difficulty and to include thinking both about the moon's shape and its apparent movement. We asked students to work in groups because we knew they would be successful and learn from each other in the process.

Evaluate

As we neared the end of our moon study, we wanted to assess student understanding in a summative manner and help students think about how far their understanding had progressed. We returned their moon phase explanations from the introductory lesson of the unit. We then asked students to write a final moon reflection, using the following prompts for guidance:

- How have your ideas about phases of the moon changed since our first moon lesson? Feel free to use diagrams to explain your thinking.

- What evidence do you have from the moon study to support your explanation?

- How would you explain the third-quarter moon to your parents (account for shape, rise time, and set time)?

Students completed this assignment individually and we evaluated their products using a scoring rubric. The rubric defined levels of understanding about phases of the moon, including the relative position and movement of the heavenly bodies. It also included a category for evidence used. There was a range of performance on the essay. All students knew that the moon's light was reflected from the sun and expressed an understanding of how the moon's lit portions were due to the relative positions of the earth, the moon, and the sun. Fewer students understood the relation of the moon's phase to its rise and set times. However, all students recognized that their ideas had changed because of the moon study. For example, Matt wrote, "I had some crazy ideas at the start, but now I really think I get it."

Note

1 This moon model is illustrated in Foster (1996).

Lessons Learned and Stories Untold

6

The vignettes in this book are based on real-life stories of science classrooms in action. The teachers who told these stories struggled with how to plan, carry out, and assess their science instruction. Their students progressed in their understanding and became more interested in science. Many of them developed deep understanding of the concepts in the units and could use these concepts to solve new problems. Others gained an initial appreciation of the ideas but will need further experiences in order to build a more complete understanding. Seamless assessment allowed us, as teachers, to form a rich picture of student understanding throughout these units.

We used the 5E model to plan and carry out science instruction linked to the assessment of student learning. We also used the 5E model to organize our telling of the vignettes. Writing these classroom stories allowed us to think about our practice in new ways. In the process, we learned quite a bit about science teaching and learning. In this final chapter, we share some of the lessons we have learned and discuss some of the stories that remain untold.

■ Lessons Learned

Lesson 1: Finding the focus concept is critical to success. In planning our 5E units, we employed the *National Science Education Standards*, our state science standards, and district curriculum guides to define learning goals that would steer instruction and assessment. The best units we developed were directed at a few big ideas, instead of addressing a laundry list of boldfaced words, as is typical in many science textbooks. Finding the focus concept for each unit required us to think deeply about science content and make decisions about what was most important to teach. In the process of concentrating on the big ideas, we sometimes had to cut out associated topics.

One of the findings of the Third International Mathematics and Science Study (TIMSS) was that the science curriculum in the United States, as compared with higher-performing countries, is a mile wide and an inch deep (Peak 1996). In contrast, our planning around a focus concept involves a less-is-more approach (American Association for the Advancement of Science 1993). The less-is-more approach contrasts with more common approaches to science teaching, such as didactic and activity-driven approaches (Anderson and Smith 1987). In didactic approaches, teachers transfer information to students and students passively receive and later retrieve information. In activity-driven approaches, teachers plan instruction around sets of activities, often for motivational, rather than learning purposes. Our seamless assessment approach is guided by what we want students to understand, and organized around the focus concept, instead of around a list of terms in a textbook. Activities are purposefully selected to support the development of conceptual understanding.

Lesson 2: The 5E is a useful tool for planning and assessing instruction. The 5E model, an expanded version of the learning cycle (Beisenherz, Dantonio, and Richardson 2000), is a research-based approach to effective science teaching. Researchers have demonstrated that, when the explanation of concepts follows firsthand experience with phenomena, students learn better. Researchers also have demonstrated that asking students to apply ideas to new problems following the explanation helps them develop the ideas more completely (Abraham and Renner 1986; Renner, Abraham, and Birnie 1988). Thus, although some of our vignettes combine two of the stages in the same lesson, the order of the stages is inviolate—exploring comes before explaining, which comes before elaborating (applying).

Using the 5E model to plan instruction helped us maintain a focus on a few major ideas (see Lesson 1). We found the 5E model to be a natural way to think about and carry out instruction: (1) hook the students, (2) explore phenomena, (3) explain phenomena, (4) use explanations to solve new problems, and (5) evaluate understanding. Connecting science concepts to the real world (especially during the *Engage* and *Elaborate* phases) improves student interest and understanding.

We also learned that tying assessment to the stages of the 5E model made sense. We were able to use assessment strategies seamlessly throughout instruction for different assessment purposes (see Table 2-2). Sometimes the same assessment strategy can be used at several different 5E stages (for example, using a concept map during *Engage* and *Evaluate* and using science notebooks at every stage). Other assessment strategies are more appropriate at specific phases (e.g., a demo memo at *Explore*, an exit ticket at *Explain*, or a final reflection at *Evaluate*). In every case, the assessment fulfills a specific purpose associated with the 5Es (see Table 2-2).

Lesson 3: Planning for assessment simultaneously with planning for instruction improves both. In our past teaching of science, we often planned science lessons and units without regard to assessment. Then, at the end of the unit, we would give a test and see how well students performed. We were good teachers who cared deeply about students' science learning, but something was missing. We discovered that that "something" was seamless assessment. Now, when we plan instruction, we are forever mindful of the learning goals we hold for students and how we can determine the degree to which students have achieved those learning goals. Knowing the learning goals helps us keep the instruction concentrated on big ideas, not factoids. Knowing what students understand provides continuous feedback that we can act on in the moment. We do not have to wait for the end of a unit to find out what students know. Instead we assess understanding throughout and get constant feedback about what students do and do not understand.

Planning for instruction and assessment simultaneously also changes our notions of what counts as evidence of learning (see Chapter 1). Using the right term or selecting the correct answer is not complete evidence of learning. Our expectations for student understanding change over the course of a unit of instruction. For example, at the beginning of a unit, we expect students to have only a rudimentary understanding or to harbor common misconceptions about the target concept. During the *Explain* phase, we expect that students are beginning to understand the concept but need more experiences with phenomena and ideas. During *Evaluate*, we expect they have achieved the learning goals. When we plan for seamless assessment in concert with instruction, we can track student understanding in light of these changing expectations. Building instruction that incorporates assessment enables teachers to stay in touch with students' conceptual development. We believe that inquiry-based instruction and seamless assessment go hand in hand in making science more meaningful for students.

Lesson 4: Seamless assessment is valuable to teachers. Looking at assessment in new ways has led to changes in our teaching practice. For example, instead of starting with definitions or formulas, we are more apt to start with *Engage* and

Explore activities. Instead of having students take notes when we talk, we are more apt to get students to talk with each other. Instead of testing through multiple-choice and matching items, we are more apt to use constructed-response and application-type questions. Monitoring students' science ideas as the instruction unfolds enables us to make instructional decisions in an organic and flexible manner. We are not chained to our lesson plans; we can respond to student needs as they arise because we are aware of them.

Seamless assessment has provided us with more complete data about student learning. Using multiple forms of assessment throughout a unit lets us see how students are thinking, what they understand, and what they need to understand. Using seamless assessment, we have developed more complete profiles of student learning. Now, when parent conference time rolls around, teachers and students have rich stories to tell parents about student science learning. When administrators ask for data on student learning, we can present them with detailed examples. Our profiles allow us to communicate with students, with their parents, and with administrators about the learning that is occurring in our science classrooms.

Lesson 5: Seamless assessment is valuable to students. In our classrooms, as in yours, learners come in all shapes and sizes and respond to our science teaching in many different ways. Through seamless assessment, we are able to tap into the variety of learning styles our students display. We have found that presenting students with opportunities to represent their learning in multiple ways—by drawing, talking, performing, solving equations, and writing—provides us with greater insight into their thinking. These opportunities to represent learning are also opportunities for students to learn. Those who typically do not succeed on end-of-unit tests often surprise us with the depth of their understanding as evidenced through seamless assessment.

Another way that seamless assessment benefits students is through illustrating to them how their learning is developing. Seamless assessment helps students be metacognitive about their learning; they think about what they do and do not understand about the concepts under study. Seamless assessment can help students recognize how they are building understanding over time. They can see that their ideas at the end of a unit of study are different from their incoming ideas. Students who are metacognitive know how to learn and learn more effectively.

Lesson 6: Collaborating with other teachers and reflecting on instruction can improve our teaching and our students' learning. It is hard to make the transition from a curriculum chock-full of learning goals to a unit focused on a few major concepts. We risk failing to cover the curriculum and not teaching all of the ideas that might be on the standardized test. It is hard to teach in ways that open up classrooms to student activity and talk. We risk noisy classrooms that may look chaotic to an outside observer. It is hard to encourage student reasoning over memorization. We risk not having all the answers. It is hard to implement inquiry-based approaches to science teaching. We risk not knowing where the students will take us. It is hard to seamlessly assess student learning throughout a unit of instruction. We risk finding out that students do not understand.

Yet when we change our practice to involve students in inquiry-based science and assess what happens, we find the benefits more than compensate for these risks. Working with other teachers to plan and reflecting with other teachers after instruction helped us confront the risks, deal with the challenges, and maximize the opportunities to improve our teaching and our students' learning. Brainstorming with other teachers generated better ideas than any of us could create on his or her own. In the community of learners we created around the notion of seamless assessment, we found time to be social and get to know each other as individuals at the same time that we enriched our professional lives.

■ Untold Stories

The vignettes in this book represent but a small fraction of our corpus of science teaching. The vignette authors have taught many science units, addressing many concepts, throughout their careers. There is a story for each of them. As we learn more about seamless assessment through the 5E model of science instruction, we anticipate that some of the units we taught in the past will change, incorporating our new learning. We also anticipate the creation of new units built around the ideals of seamless assessment. Clearly, there are many stories yet to tell.

Secondly, although we have tried to represent student voices in our vignettes, it was necessary to tell our stories from the teachers' perspectives. What if we were to ask students to tell their science learning stories? Would we hear something different? What if we asked parents their views of their children's science learning? What would we hear? And what if we asked administrators for their perspectives? We hope you can make the effort to seek out these other stories as you pursue seamless assessment in your teaching.

We realize this book is not a complete guide to elementary and middle school science assessment. We never intended it to be. For example, we have not included much about the considerable research that demonstrates the value of inquiry-based instruction and seamless assessment. Furthermore, we have illustrated only a small part of teachers' assessment work. Besides planning and enacting seamless assessment, the teachers in this book were also accountable for scoring and grading student work, examining student growth over time, and communicating with administrators and parents about student learning. To fulfill these responsibilities, the teachers developed scoring rubrics, student work portfolios, grade cards, student-led parent conference forms, and other tools. Yet none of these activities would have been possible without first planning and carrying out inquiry-based instruction accompanied by seamless assessment. We leave it to other books to help you think about issues related to grading and the like. This book stands as a guide to implementing seamless assessment in your science classroom.

We hope that the vignettes in this book will motivate you to use inquiry-based instruction and seamless assessment to guide your journeys in science teaching. We encourage you to use a less-is-more approach and to introduce explanations after experiences. We support your progress in assessing students throughout the 5 Es of science instruction. As a result, we expect that your students will have increased opportunities to learn science as they do, talk, and think science. We wish you the best and hope to hear the stories of your travels.

Online Resources for Assessment Strategies

The websites we selected provide how- to instructions and examples of application in educational settings for many of the assessment strategies listed in Chapter 2. They represent a fraction of those available but offer some good starting points.

STRATEGY	SOURCE	INTERNET ADDRESS
Application Problem	NISE—National Institute for Science Education (University of Wisconsin, Madison)	www.wcer.wisc.edu/archive/cl1/CL/doingcl/prbsolv.htm
Brainstorming	NISE	www.wcer.wisc.edu/archive/cl1/CL/doingcl/brain.htm
Card Sort Task	*The Shape of Life* (PBS) *Activity Guide*	www.pbs.org/kcet/shapeoflife/resources/activity4.pdf
Comparison Essay	The Langara Writing Center (Langara College, British Columbia, Canada)	www.langara.bc.ca/writingcentre/handouts /comparisonessay.html
Concept Mapping	*Field-Tested Learning Assessment Guide* (FLAG), NISE	www.wcer.wisc.edu/archive/cl1/flag/cat/catframe.htm
concepTest	FLAG, NISE	www.wcer.wisc.edu/archive/cl1/flag/cat/catframe.htm
Conceptual Cartoon	Angel Solutions Limited	www.conceptcartoons.com/what_is_a_concept_cartoon.html
Constructed Response	edteck	www.edteck.com/dbq/testing/const_resp.htm
Debate	Debate Central (University of Vermont)	http://debate.uvm.edu/default.html
Design Activity	Science NetLinks—American Association for the Advancement of Science	www.sciencenetlinks.org/sci_update_index.cfm
Discrepant Event	*Invitations to Inquiry*, by Tik Liem	www.scienceinquiry.com/demo-main.htm
Drawing Completion	The Physics Classroom— Mathsoft Engineering and Education	www.physicsclassroom.com/Class/newtlaws/u2l2c.html
Exit Sheet, Exit Ticket, and Minute Paper	NISE	www.wcer.wisc.edu/archive/cl1/CL/doingcl/qkwrite.htm

STRATEGY	SOURCE	INTERNET ADDRESS
Final (Essay) Reflection	Texas A&M University Writing Center	http://writingcenter.tamu.edu/content/view/61/0/
Interview	FLAG, NISE	www.wcer.wisc.edu/archive/cl1/flag/cat/catframe.htm
KWL Chart	North Central Regional Educational Laboratory	www.ncrel.org/sdrs/areas/issues/students/learning/lr2kwl.htm
Making Models	Boston Museum of Science	www.mos.org/exhibitdevelopment/skills/models.html
Memoir	Kentucky Department of Education	www.ftcpublishing.com/rwiky.pdf
One-Page Memo	*Learning Marketing Principles Through Memo Writing*, Thomson South-Western	www.swlearning.com/marketing/gitm/gitm25-6.html
Pair Problem Solving	NISE	www.wcer.wisc.edu/archive/cl1/CL/doingcl/tapps.htm
Poster	Information Literacy, Cape Higher Education Consortium	www.lib.uct.ac.za/infolit/poster.htm
Predicting Activities	Environmental Protection Agency: New England, Project A. I. R. E. (Air Information Resources for Educators) *Warm-Up Exercise: Prediction*	www.epa.gov/NE/students/pdfs/warm_a.pdf
Presentation	NISE	www.wcer.wisc.edu/archive/cl1/CL/doingcl/reports.htm
Puzzlers	NOVA Online (PBS)	www.pbs.org/wgbh/nova/tothemoon/puzzlers.html
Questionnaire	FLAG, NISE	www.wcer.wisc.edu/archive/cl1/flag/cat/catframe.htm
Science Notebooks	Elementary Science Integration Projects, University of Missouri-St. Louis	www.esiponline.org/classroom/foundations/writing/notebook.html
Self-Evaluation	TeacherVision.com	www.teachervision.com/tv/printables/0130533688_ALFL0402.pdf
Think, Pair, Share	NISE	www.wcer.wisc.edu/archive/cl1/CL/doingcl/thinkps.htm
Thought Experiment	*Stanford Encyclopedia of Philosophy*	http://plato.stanford.edu/entries/thought-experiment/
Venn Diagrams and other graphic organizers	Education Place	www.eduplace.com/graphicorganizer/

References

ABELL, SANDRA, GAIL ANDERSON, and JANICE CHEZEM. 2000. "Science as Argument and Explanation: Inquiring into Concepts of Sound in Third Grade." In *Inquiring into Inquiry Learning and Teaching in Science*, ed. Jim Minstrell and Emily van Zee, 65–79. Washington, DC: American Association for the Advancement of Science.

ABELL, SANDRA, MELISSA GEORGE, and MARIANA MARTINI. 2002. "The Moon Investigation: Instructional Strategies for Elementary Science Methods." *Journal of Science Teacher Education, 13*: 85–100.

ABRAHAM, MICHAEL R., and JOHN W. RENNER. 1986. "The Sequence of Learning Cycle Activities in High School Chemistry." *Journal of Research in Science Teaching, 23*: 21–43.

ALLEN, PAMELA. 1982. *Who Sank the Boat?* New York: Penguin Putnam.

AMERICAN ASSOCIATION FOR THE ADVANCEMENT OF SCIENCE. 1993. *Benchmarks for Science Literacy*. New York: Oxford University Press.

ANDERSON, CHARLES W., and EDWARD L. SMITH. 1987. "Teaching Science." In *Educators' Handbook: A Research Perspective*, ed. Virginia Richardson-Koehler, 84–111. New York: Longman.

ANDERSON, LORIN, and DAVID KRATHWOHL, eds. 2001. *A Taxonomy for Learning, Teaching, and Assessing: A Revision of Bloom's Taxonomy of Educational Objectives*. New York: Longman.

BEISENHERZ, PAUL C., MARYLOU DANTONIO, and LON RICHARDSON. 2000. "The Learning Cycle." *Science Scope, 24* (4): 34–38.

BERLINER, DAVID C. 1994. "Expertise: The Wonder of Exemplary Performances." In *Creating Powerful Thinking in Teachers and Students*, ed. John N. Mangieri and Cathy Collins Block, 161–86. Fort Worth, TX: Holt, Rinehart and Winston.

BLACK, PAUL. 2003. "The Importance of Everyday Assessment." In *Everyday Assessment in the Science Classroom*, ed. J. Myron Atkin and Janet E. Coffey, 1–11. Arlington, VA: NSTA Press.

BRANSFORD, JOHN, ANN BROWN, and RODNEY COCKING, eds. 1999. *How People Learn: Brain, Mind, Experience, and School*. Washington, DC: National Academy Press.

BYBEE, RODGER. 1997. *Achieving Scientific Literacy: From Purposes to Practices*. Portsmouth, NH: Heinemann.

————. 2002. "Scientific Inquiry, Student Learning, and the Science Curriculum." In *Learning Science and the Science of Learning*, ed. Rodger Bybee, 25–35. Arlington, VA: NSTA Press.

CAMPBELL, BRIAN, and LORI FULTON. 2003. *Science Notebooks: Writing About Inquiry*. Portsmouth, NH: Heinemann.

CARLE, ERIC. 1990. *The Very Quiet Cricket*. New York: Philomel.

CONKLIN, TOM. 2000. *Disasters!* New York: Scholastic.

DORAN, RODNEY, FRED CHAN, PINCHAS TAMIR, and CAROL LENHARDT. 2002. *Science Educators' Guide to Laboratory Assessment*. Arlington, VA: NSTA Press.

DRIVER, ROSALIND, ANN SQUIRES, PETER RUSHWORTH, and VALERIE WOOD-ROBINSON. 1994. *Making Sense of Secondary Science: Research into Children's Ideas*. New York: Routledge.

FOSTER, GERALD. 1996. "Look to the Moon." *Science and Children, 34* (3): 30–33.

FRIEDRICHSEN, PATRICIA M., and THOMAS M. DANA. 2002. "Using a Card-Sorting Task to Elicit and Clarify Science Teaching Orientations." *Journal of Science Teacher Education, 14*: 291–309.

GEORGE, JEAN CRAIGHEAD. 1971. *Who Really Killed Cock Robin?* New York: Harper Trophy.

————. 1992. *The Missing 'Gator of Gumbo Limbo*. New York: Harper Trophy.

GEORGE, MELISSA. 2005. "Looking Inward, Looking Outward: Developing Knowledge Through Teacher Research in a Middle School Science Classroom During a Unit on Magnetism and Electricity." Doctoral Dissertation, Purdue University, West Lafayette, IN.

GRIBBIN, JOHN. 2002. *The Scientists: A History of Science Told Through the Lives of Its Greatest Inventors*. New York: Random House.

HAKIM, JOY. 2004. *The Story of Science, Book One: Aristotle Leads the Way*. Washington, DC: Smithsonian Books.

HEIN, GEORGE, and SABRA PRICE. 1994. *Active Assessment for Active Science: A Guide for Elementary School Teachers*. Portsmouth, NH: Heinemann.

HILL, DEREK L. 2004. *Latin America Shows Rapid Rise in S and E Articles*. (NSF 04-336). Washington, DC: National Science Foundation Directorate for Social, Behavioral, and Economic Sciences.

JELLY, SHEILA. 2001. "Helping Children Raise Questions—and Answer Them." In *Primary Science: Taking the Plunge*, 2d ed., ed. Wynne Harlen, 36–47. Portsmouth, NH: Heinemann.

JOHNSON, CROCKETT. 1981. *Harold and the Purple Crayon*. New York: HarperCollins.

JOHNSON, ROGER T., and DAVID W. JOHNSON. 1985. "Using Structured Controversy in Science Classrooms." In *Science Technology Society* (1985 Yearbook of the National Science Teachers Association), ed. Rodger W. Bybee, 228–34. Washington, DC: National Science Teachers Association.

KEHRET, PEG. 1998. *The Volcano Disaster*. New York: Pocket.

LIEM, TIK L. 1987. *Invitations to Science Inquiry*. 2d ed. Chino Hills, CA: Science Inquiry Enterprises.

MAZUR, ERIC. 1997. *Peer Instruction: A User's Manual*. Upper Saddle River, NJ: Prentice Hall.

NATIONAL ACADEMY OF SCIENCES (WORKING GROUP ON TEACHING EVOLUTION). 1998. *Teaching About Evolution and the Nature of Science*. Washington, DC: National Academy Press.

NATIONAL RESEARCH COUNCIL. 1996. *National Science Education Standards*. Washington, DC: National Academy Press.

——— . 2000. *Inquiry and the National Science Education Standards*. Washington, DC: National Academy Press.

NATIONAL SCIENCE RESOURCES CENTER. 1997. *Land and Water*. Science and Technology for Children. Burlington, NC: Carolina Biological Supply.

——— . 2000. *Catastrophic Events*. Science and Technology Concepts for Middle Schools (STC/MS). Burlington, NC: Carolina Biological Supply.

——— . 2002. *Soils*. Science and Technology for Children. Burlington, NC: Carolina Biological Supply.

NOVAK, JOSEPH D. 1998. *Learning, Creating, and Using Knowledge: Concept Maps as Facilitative Tools in Schools and Corporations*. Mahwah, NJ: Lawrence Erlbaum.

NOVAK, JOSEPH D., and D. BOB GOWIN. 1984. *Learning How to Learn*. New York: Cambridge University Press.

OGLE, DONNA. 1986. "A Teaching Model That Develops Active Reading of Expository Text." *The Reading Teacher, 39*: 564–70.

OSBORNE, ROGER, and PETER FREYBERG. 1985. *Learning in Science: The Implications of Children's Science*. Portsmouth, NH: Heinemann.

PEAK, LOIS. 1996. *Pursuing Excellence: A Study of U.S. Eighth-Grade Mathematics and Science Teaching, Learning, Curriculum, and Achievement in International Context: Initial Findings from the Third International Mathematics and Science Study*. Washington, DC: National Center for Education Statistics.

PESTEL, BEVERLY C. 1993. "Teaching Problem Solving Without Modeling Through 'Thinking Aloud Pair Problem Solving.'" *Science Education, 77*: 83–94.

PROJECT WET WATER EDUCATION FOR TEACHERS. 1995. *Project WET Curriculum and Activity Guide*. Bozeman, MT: Author.

REARDON, JEANNE. 1993. "Developing a Community of Scientists." In *Science Workshop: A Whole Language Approach*, ed. Wendy Saul, Jeanne Reardon, Anne Schmidt, Charles Pearce, Dana Blackwood, and Mary Dickinson Bird, 19–38. Portsmouth, NH: Heinemann.

RENNER, JOHN W., MICHAEL R. ABRAHAM, and HOWARD H. BIRNIE. 1988. "The Necessity of Each Phase of the Learning Cycle in Teaching High School Physics." *Journal of Research in Science Teaching, 25*: 39–58.

SMITH, WALTER. 2003. "Meeting the Moon from a Global Perspective." *Science Scope, 26* (8): 24–28.

VICTOR, EDWARD, and RICHARD D. KELLOUGH. 2000. *Science for the Elementary and Middle School*. Upper Saddle River, NJ: Merrill.

VOLKMANN, MARK, and SANDRA ABELL. 2003. "Seamless Assessment." *Science and Children, 40* (8): 41–45.

WANDERSEE, JAMES, JOEL MINTZES, and JOSEPH NOVAK. 1994. "Research on Alternative Conceptions in Science." In *Handbook of Research on Science Teaching and Learning*, ed. Dorothy Gabel, 177–210. New York: Macmillan.

WILSON, MARK, and KATHLEEN SCALISE. 2003. "Reporting Progress to Parents and Others: Beyond Grades." In *Everyday Assessment in the Science Classroom*, ed. J. Myron Atkin and Janet E. Coffey, 89–108. Arlington, VA: NSTA Press.

ZIKE, DINAH. 2001. *Big Book of Science for Middle School and High School*. San Antonio, TX: Dinah-Might Adventures.

About the Authors

■ Authors

Sandra K. Abell is Professor of Science Education and Director of the Science Education Center at the University of Missouri-Columbia. A former elementary school science teacher, Abell has conducted numerous teaching and research projects in elementary and middle school science classrooms in collaboration with classroom teachers. Her science teaching ideas have appeared in *Science and Children*, *Science Scope*, *The Science Teacher*, and *Science Activities*. Her research, focused on science teacher learning, has been published in journals such as *Science Education*, *Journal of Research in Science Teaching*, *Journal of Science Teacher Education*, and *International Journal of Science Education*. She is coeditor of the forthcoming *Handbook of Research on Science Education* and a past president of the National Association for Research in Science Teaching (NARST).

Mark J. Volkmann is Associate Professor of Science Education at the University of Missouri-Columbia. A former junior high and high school science teacher, Volkmann has collaborated with teachers and scientists on numerous projects in middle school science classrooms and in science teacher education. His science teaching ideas have appeared in *Science and Children*, *The Science Teacher*, and *School Science and Mathematics*. His research, focused on science teacher identity and teaching science through inquiry, has been published in *Science Education*, *Journal of Research in Science Teaching*, and *Journal of Science Teacher Education*.

■ Contributors

The teachers who wrote vignettes of science classrooms in action for this volume teach for Columbia Public Schools (CPS) in Columbia, Missouri. CPS enrolls more than sixteen thousand students in grades K–12. The district includes nineteen elementary (K–5) buildings, three middle schools (grades 6–7), three junior highs (8–9), and three high schools (10–12). In CPS, elementary science is taught by classroom teachers in grades K–3 and by intermediate science specialists beginning in grade 4. Science teachers form part of each instructional team at the middle school level and work in science departments in junior high schools. The teachers who contributed to this book held the following positions in CPS at the time of writing:

- **Julie Alexander,** intermediate science specialist, Ridgeway Elementary and Paxton Keeley Elementary

- **Cathy Carpenter,** science teacher, Lange Middle School

- **Tracy Hager,** third-grade teacher, Shepard Boulevard Elementary

- **Meera Sood,** science teacher, Smithton Middle School

- **Sara Torres,** science coordinator, Columbia Public Schools (and former middle-level science teacher for the district)

- **Kelly Turnbough,** science teacher, Lange Middle School

- **Marsha Tyson,** science teacher, Oakland Junior High

- **Kenneth Welty,** second-grade teacher, Paxton Keeley Elementary

- **Laura Zinszer,** intermediate science specialist, Blue Ridge Elementary, Parkade Elementary, and Ridgeway Elementary

One vignette was contributed by **Michele Lee,** a doctoral student in science education at the University of Missouri-Columbia and a former elementary teacher in Maryland and Massachusetts.